VGM Opportunities Series

OPPORTUNITIES IN **DRAFTING CAREERS**

Mark Rowh

Foreword by

Rachel H. Howard
Executive Director
American Design Drafting Association (DDA)

VGM Career Horizons
a division of *NTC Publishing Group*
Lincolnwood, Illinois USA

Cover Photo Credits:
Clockwise from top left: Photo Network; Signal VOP Group; Autodesk; VGM.

Library of Congress Cataloging-in-Publication Data

Rowh, Mark
Opportunities in drafting careers / Mark Rowh.

p. cm. — (VGM opportunities series)
ISBN 0-8442-4082-6 — ISBN 0-8442-4083-4 (pbk.)
1. Mechanical drawing—Vocational gudiance. I. Title
II. Series.

T356.R68 1993
604.2′023—dc20 93-10586
CIP

Published by VGM Career Horizons, a division of NTC Publishing Group.

Manufactured in the United States of America.

3 4 5 6 7 8 9 0 VP 9 8 7 6 5 4 3 2 1

ABOUT THE AUTHOR

Mark Rowh serves on the administrative staff at New River Community College in Dublin, Virginia. He previously served in administrative positions at Bluefield State College, Parkersburg Community College, and Greenville Technical College, where his duties included working with various technical programs.

A graduate of West Virginia State College and Marshall University, he has also completed a doctoral degree in vocational and technical education at Clemson University.

Rowh is a widely published free-lance writer, with work in more than one hundred magazines. He has written a number of books, including several volumes in the VGM Career Horizons series.

ACKNOWLEDGMENT

The author wishes to acknowledge the contribution of Benjamin Stern, author of earlier editions of this book, and to extend thanks for the inclusion of the valuable material from the earlier works.

FOREWORD

As you consider a drafting career, do not lose sight of the valuable service you will perform. Not one thing can be produced unless the drafter has made a graphic or pictorial representation of the item first. Drafters fill the vital link of assuring communication between the individual with the idea and the individual who actually manufactures the product.

The advent of the computer revolutionized the way drafters operate. With computer assisted drafting (CAD), a drafter can explore many different views before committing a drawing to paper. Evolving technology will continue to change the profession. An example is the use of "virtual reality" programs to simulate actually being within or beside the object the drafter is working on. Technology such as this will add new dimensions to the profession.

The opportunities for drafters are varied—architectural, electronics, construction, mechanical, landscape, utilities, and educational institutions all seek qualified individuals. A drafting career is exact and demanding but offers rewards in knowing you are doing a task which is required to complete the link between the designer and the user of the product.

I know you will find the profession rewarding and wish you the very best.

Rachel H. Howard
Executive Director
American Design Drafting Association (ADDA)

DEDICATION

To Linda, Lisa, David, and Jenny

CONTENTS

CHAPTER 1

THE SCOPE OF DRAFTING

> Drafting (draf' ting) n. The systematic representation and dimensional specification of mechanical and architectural structures.
>
> —*The American Heritage Dictionary*

If you want to see the work of a drafter, you will not have to go far. Just tune in the radio, flick on the living room lights, or go for a drive in a modern car. All around you are the manufactured devices and structures that are the hallmark of our technological age. Each structure represents the work of many people, among them those who prepared the drawings on which its construction was based: *the drafters*.

Just what do drafters do? They produce very detailed drawings of structures or objects. These precise drawings show the exact size, shape, dimensions, and other specifications of the object, including all its components. Once the drawings are made, they become the plans that other workers use to build the planned structure or device.

As an example, think of America's newest space shuttle, *Endeavour.* As the huge space vehicle descends out of the blue sky and makes a perfect desert landing, it is hard to imagine a more magnificent piece of machinery. Perhaps the epitome of modern technological advancement, the shuttle can withstand the tremendous force of a rocket launching, the hostile environment of airless space, and the intense heat of reentry into the earth's atmosphere. As impressive as it is, the versatile spacecraft would not even exist if not for the work of drafters. For each of the thousands of components making up the shuttle had to be designed and built for a specific purpose, and it was the drawings of drafters that made the construction possible.

Spacecraft, computers, airplanes, and other glamorous products of our technological age are just the latest in a long, continuous stream of developing technology, with drafting playing a key role along the way. As early as 2400 B.C., Sumerian builders such as Gudea, ruler of the city-state of Lagash, used drafting techniques in designing temples and other public buildings. Using soft clay tablets and a stylus, they began a tradition that has extended on through the present, a period of more than 4,400 years.

Through all these years drafters have practiced their craft wherever there existed a civilization that built roads, bridges, buildings, and the like, or that constructed machinery, devices, and tools. As technology grew, so did the stature of drafters, until now they are key figures in the industrial and cultural life of the world. They can be found in every branch of manufacturing, maintenance, construction, design, military materials, science, and government, or wherever that universal language known as mechanical drawing is used.

It is the drafters' mastery of this graphic language that has enabled them to attain their status. With this language, communication with all people becomes possible, though they may not speak the same tongue. By the use of these mechanical drawings or pictorial representations, inhabitants of different countries and districts, using different languages or dialects, may still talk to one another and be understood. Almost all of us have had the experience of putting together a toy or assembling a model or a piece of furniture or some utensil, using as a guide the diagrams or schematic drawings that usually come with the unassembled object. Even though most people have not had technical training, they can still follow these drawings and assemble the object. Most of us can use a road map or understand simple architectural drawings and other diagrams that often illustrate newspaper stories and the like. When we do this, we are following mechanical drawings in a very simplified form.

The drafter, an expert in the use of this universal system of pictorial representation, stands in the middle as the translator, the transmitter or connection, between the idea and the accomplishment of the idea. Taking the rough notes, instructions, calculations, and sketches of the engineer, the designer, the architect, the scientist, the inventor, and others, the drafter translates them into systematic drawings. These drawings are understood, followed, and used by the construction worker, the machinist, and the person in the shop to produce the machine, the building, the airplane, the bridge, the road, the satellite, the missile, or any of the multitudinous artifacts of our marvelous age of science and technology. The drafter combines technical knowledge with practical skills to help bring ideas to reality.

Typical drafting roles, as noted by Tennessee's Northeast State Technical College in its catalog, include the following:

Architectural Drafter - Draws architectural details for the development of a construction site. The drafter analyzes architectural sketches and develops working construction drawings.

Tool Designer - Designs tools, jigs, gages, fixtures; draws and designs anything dealing with manufacturing or handling of parts in the manufacturing process.

Mechanical Drafter - Specializes in making detailed working drawings of machinery and mechanical devices; drafts multiple assembly drawings as required for manufacture.

Product Designer - Develops detailed design drawings and specifications of mechanical equipment; analyzes engineering sketches and drawings to determine design factors.

Pipe Drafter - Draws and designs piping systems. Works in petrochemical, nuclear, and solar plants. The drafter may use models and computer-aided drafting machines.

Structural Detailer - Produces basic structural drawings, utilizes reference materials to prepare drawing material lists, estimates and writes specifications for total job.

Electrical/Electronics Drafter - Drafts electrical schematics, motor and control schematics, lighting panels, connections such as wiring diagrams, connection diagrams, and logic diagrams.

CAD Drafter - Drafters using computers with special software. The drafter may draw from varying drafting disciplines.

As A. H. Rau puts it in his *Drafting for Good Reproduction,* "A good drawing is easy to recognize by clear, clean line work, well selected views, legible dimensioning, and simplicity of presentation—not by flourishes or useless artistry. The importance of simplicity cannot be overemphasized as a prime factor in contributing to legible reproduction. In addition to the drafter's basic responsibility to convey engineering instructions, he must also be mindful of the requirement that his drawing must be of quality sufficient to produce multiple copies. He must also use drafting techniques and practices which will assure economical and effective reproductions."

The *Dictionary of Occupational Titles* (D.O.T.), published by the United States Department of Labor, defines the drafter's work as follows:

> Prepares working plans and detail drawings from rough or detailed sketches and notes for engineering or manufacturing purposes according to dimensional specifications: Calculates and lays out dimensions, angles, curvature of parts, materials to be used, relationship of one part to another, and relationship of various parts to entire structure or project, utilizing knowledge of engineering practices, mathematics, building materials, manufacturing technology, and related physical sciences. Creates preliminary or final sketch of proposed drawing, using standard drafting

> techniques and devices, such as drawing board, T-square, protractor, and drafting machine, or using computer-assisted design/drafting equipment. Modifies drawings as directed by engineer or architect. Classifications are made according to type of drafting, such as electrical, electronic, aeronautical, civil, mechanical, or architectural.

The *D.O.T* describes the work of a "computer assisted drafter" as one who:

> Drafts layouts, drawings and designs for application in such fields as aeronautics, architecture, or electronics, according to engineering specifications, using computer: Reviews engineering drawings and supporting documents to verify freedom of movement between parts and adherence to company or industry standard practices and adequacy of parts identification. Analyzes design and confers with engineering staff to resolve details not completely defined. Locates file relating to projection data base library and loads program into computer. Retrieves information from file and displays information on cathode-ray-tube (CRT) screen, using required computer language. Types commands to rotate or zoom-in on display to redesign, modify, or otherwise edit existing design. Traces over face of photosensitive screen to redraw details or rewrite text. Displays final drawing on screen to verify completeness, clarity and accuracy of drawing. Types command to transfer drawing dimensions from computer onto hardcopy, using peripheral equipment, such as digitizer or plotter controlled by computer. Submits completed drawings to supervisor for review.

THE NATURE OF THE DRAFTER'S WORK

Exactly how does a drafter go about her or his daily work? That all depends on whether traditional methods are followed, or the drafter is using the newer computer techniques described above. With traditional equipment (which is still used in some companies), he or she usually stands or sits before a large drawing board or table that can be adjusted to provide a comfortable working height and angle. Upon the surface of this board or table is spread and fastened, usually with drafting tape, a sheet of layout or drawing paper of the size required for the particular job, or tracing paper, tracing cloth, graph paper, or whatever sort of material is necessary for the work to be done. Or if computer equipment is used, the work will be done at a computer work station, with the drawings appearing on a video monitor during the preliminary stages, and then being reproduced by a computer printer when a paper copy is needed. Working from notes, specifications, rough or detailed sketches, previous job drawings, or any other type of instruction, the drafter makes clear, complete, and accurate assemblies, subassemblies, working drawings, and the like, according to the correct dimensions.

In other words, the drafter whether starting on a new project or working on an existing one, must proceed on the basis of information and ideas derived from the engineer, architect, builder, or project leader in charge or from existing plans, drawings, specifications, and the like, plus her or his own experience.

For example, he or she may be employed in the office of an architect who has been developing ideas, sketches, and drawings for a client who is planning to build a new office building. When the ideas and sketches have been approved, the architect or the office chief will sit down with the drafter to plan the necessary drawings. The drafter will be given all the information gathered on the project and will then proceed, probably beginning with a site plan showing the relation of the proposed building to the site upon which it is to be built, together with the information derived from the land survey. Next the drafter will draw the foundation plans, floor plans, elevations, sections, and the rest. Meanwhile, other drafters will be taking details from these plans in order to develop drawings for the plumbing, heating and ventilating, light and power, and any other feature needed to complete the building.

In like manner, a drafter employed in any other aspect of manufacturing, maintenance, or industry would proceed from information and instructions supplied by the engineer, the project leader, the chief drafter, and any existing drawings, specifications, and experience. In any case, he or she would keep a constant check on the materials to be used, the relation of one part to another, the relations of the various parts to the shapes, dimensions, and weights of various other items, and to the whole structure. He or she would perform the necessary mathematical calculations for strength of materials, sizes, tolerances, and shapes and would consult technicians or professionals as well as handbooks and technical standards for the information necessary to do the job.

The drafter is trained in the theory and principles of mechanical drawing. This is based on a systematic, mathematical approach known as descriptive geometry. Also, as the term indicates, it is a system of drawing that uses mechanical means (instruments and tools) for its production, as contrasted to the free-hand methods of the artist or illustrator. The drafter, therefore, has a whole battery of precision tools at his or her command. The basic ones used in noncomputerized drafting are the T-square and triangles, by means of which the drafter can produce straight horizontal, vertical, and slant lines at 45°, 60°, and 30°, as well as combinations of these. In many installations, the T-square and triangles have been replaced by a drafting "machine." Attached to the drawing board or table the drafting machine combines the functions of the basic tools and allows for a great deal of flexibility in producing horizontal, vertical, and slant lines as well as saving time in tool handling.

The drafter will use compasses for arcs and circles and dividers for accurate spacing. Scales (rulers) provide for accurate measurements and french or irregular curves are on hand for curved lines. For ink work he or she can use ruling pens for straight lines and bow pens and compasses for circles. The pencils used are specially made and come in various degrees of hardness as required for the job. They are kept properly sharpened; often the drafter has a special sharpening device for the purpose. There are also other instruments on hand that he or she can use for special purposes.

In addition to this equipment, modern drafting includes the computer and related "high tech" equipment such as plotters, digitizers, high resolution monitors, and various kinds of sophisticated software.

THE DRAFTER'S INSTRUMENTS

Most of the instruments just described have been in existence in one form or another for many centuries. For example, the ancient Egyptians, who were master builders, used a piece of string as a compass for laying out circles or arcs. A bronze point passing through a loop in the string was used by one hand to describe the desired circle or arc while the other hand held the string at the center of the circle. Many of us have used a similar method at one time or another when no compass was handy.

By the time of the Romans, two-legged metal compasses had been developed in a form quite similar to those used today. Specimens of these and of the ivory-wood rulers used in those ancient times can be seen in various museums. Lines were usually scratched by a sharpened metal stylus on such surfaces as wood, papyrus, or vellum; then ink was traced over these lines with a sharpened reed pen. Paper was brought into Europe from China by the Arabs in the twelfth century, at which time quill pens made from goose feathers became quite common.

One of the greatest advances for artists, drafters, architects, and others occurred when the lead pencil was invented in England. Mention of this was made in publications around the year 1565, when the pencil was described as a core of graphite encased in wooden blocks. However, because of the softness of the graphite used, pencil drawings were apt to be badly smudged. It was not until 1795 that Faber, in Germany, arrived at a practical method of controlling the hardness of the pencil.

In the museum at Mt. Vernon, Virginia, the home of George Washington, a set of drawing instruments is displayed. Washington used these instruments as a young man when he served as a land surveyor. The compasses, dividers, ruling pen, and other tools in this set bear a strikingly modern look, although their construction is much heavier than that of today's tools.

CAD: COMPUTER ASSISTED DRAFTING

The most revolutionary advancement in the history of drafting has occurred in recent years with the advent of the computer. Computer assisted drafting (CAD) has dramatically changed the way drafters operate, as the earlier description from the Department of Labor illustrates. Thanks to the computer, a drafter can explore different options without actually committing a drawing to paper, simply by entering information into the computer and then viewing the image produced on a CRT or monitor. These computer images can be rotated, viewed from any angle, magnified as needed, and even viewed in three dimensions. And if a complicated drawing needs to be altered, the changes can be made in a few minutes rather than the hours required to re-do a drawing on paper.

DUPLICATING THE DRAWING

When the drawings are finished, checked, and approved, they then are reproduced for use in the shop, the shipyard, on a construction site, in the tool-making room, in the laboratory, or wherever they are required. The duplication may be done by several methods, including blueprinting, which is the oldest and still the most widely used. The very word *blueprint* is familiar to all of us as a synonym for duplication.

The process consists of coating a white sheet or roll of paper with a chemical that is sensitive to light. The coated paper is kept away from light until it is used, at which time the drawing on translucent tracing paper or cloth is laid over it and exposed to light for the proper length of time. The light, penetrating the drawing where there are no lines or figures drawn on it, strikes the sensitized paper at these open spots and sets the chemical. After exposure the blueprint paper is washed in clear, clean water and these spots turn blue. Those places covered by the lines and figures of the drawing, and therefore not affected by the light, are washed clean. The result is a duplicated drawing of white lines and figures against a blue background.

Machines, some quite elaborate, have been developed to handle the operations necessary for drawing reproductions. All that is necessary in their use is for the operator to feed the drawings into the machine. Many people started careers in the drawing room as operators of such machines.

CLASSIFICATION OF DRAFTERS

Drafters are usually classified as technicians or semiprofessionals. Important as they are, and although they usually have a technical education and background, they are *not* engineers. A drafter need not have a technical degree or the years of training necessary to be an engineer. In fact, it is usually a waste of valuable manpower to use an engineer on the drafting board, unless that engineer is a specialist in advanced design. Even then, the engineer usually needs only to produce sketches and plans to be worked up and finished by drafters.

An industrial survey once made of metal trades in the New York metropolitan area included a special section on drafters. It was interesting to note that in practically every plant sampled, the opinion was that a good three-year vocational-technical course including technical drafting was an excellent start for a young drafter. Practically all the chief drafters, foremen, and others concerned with the hiring and direction of drafters concurred with this and were of the belief that an engineering degree was not needed. In fact, the opinion was expressed that such a degree was often a handicap because the engineer is not specifically trained for drafting work, is over-educated for this purpose, and will not stay any longer at the drawing board than can be helped. The feeling among these leaders was that the drafter is a highly skilled technician or semi-professional. It is true, of course, that some engineers are forced by circumstances to work as drafters, and in such branches of the industry as ship design, architectural design, or structural design, the top designers and leaders often do have advanced technical training.

However, while high school or vocational school training or an apprenticeship may be enough to start a young drafter on a chosen career, the ever-increasing complexities of our technologies make it imperative that the beginner continue her or his technical education after high school or apprenticeship. Fortunately many two-year colleges and technical institutes have been established in all states over the last thirty years. These offer terminal two-year courses leading to an associate degree in practically any technology. In most of these schools, courses can be taken after working hours.

Because such associate degrees are available, it is unnecessary for the young drafter to take an engineering degree to qualify as a designer. He or she can take advantage of two-year programs to perfect her or his grasp of the technology of a particular branch of the craft. In addition, if the young drafter does plan to go on to become an engineer or an architect, he or she can get a good start in these two-year courses. This is discussed in more detail in a later chapter.

A list of colleges and technical institutes offering two-year programs can be found in Appendix E at the back of the book. These schools offer courses in the technologies that affect drafters.

WOMEN IN DRAFTING

Women have always been important members of the labor force. Their numbers depend upon various social, economic, political, and industrial conditions. In the past, women operators were in the great majority in textile-weaving and fiber-spinning plants; the health service fields were dominated by women; and domestic servants were mostly women. But very few of them managed to become skilled in crafts such as those of the carpenter, the machinist, the mechanic and the like until World War I. Personnel shortages then forced industry to train women as welders, riveters, machine operators, shipfitters, and practitioners of other crafts. Between World War I and World War II, however, women were phased out of the skilled crafts. The onset of World War II brought them back again as men were drafted by the millions into the armed services. Once again, between 1941 and 1945, women entered the war industries as machine operators, welders, and drafters. That they did a good job was acknowledged, as in this item from *The American Machinist* of September 17, 1942:

> Industrialists are discovering that women are thoroughly competent to do work formerly reserved for men. Some manufacturers would whisper to you, in confidence, that women after a relatively short training period, are producing more and better than men ever did.

During World War II, 6.7 million women entered the labor force. Of these, 2.9 million were first-time entrants into the category of craftswomen and operators. Again, as after World War I, when the emergency ended, most of the women were gradually eased out of the skilled jobs.

In the 1960s, the rapidly escalating movement toward ending job discrimination on the basis of sex, race, or creed resulted in federal enactment of the 1964 Civil Rights Act. Among other things, this legislation assured equal opportunities for women to enter any occupation. As a result, the number of women employed as craft workers has increased steadily over the years, with over one-half million women now employed. The Civil Rights Act is enforced by a Federal Equal Employment Opportunity Commission.

Every effort is now being made by this commission to enforce the law. For example, The Women's Bureau of the United States Department of Labor published a bulletin called "A Working Woman's Guide to the Job Rights" in which the following suggestions were listed.

You have a right to complain if:

- an employer's advertisement for employees carries a sex label;
- an employer refuses to let you file an application but accepts others;
- a union or employment agency refuses to refer you to job openings;
- a union refuses to accept you into membership;
- you are fired or laid off without cause;
- you are passed over for promotion for which you are qualified;
- you are paid less than others for comparable work;
- you are placed in a segregated seniority line;
- you are left out of training or apprenticeship programs;
- the reason for any of these acts is your sex, race, color, religion, or national origin;
- your employer provides racially segregated lunchrooms, locker rooms, restrooms, or recreational facilities.

The bulletin suggested that complaints be filed with the nearest office of the Federal Equal Employment Opportunity Commission. Over the years since the passage of the Civil Rights Acts, many lawsuits based upon these complaints have been brought to court and tried, with quite a number of successful results for the women involved. These results have caused employers to realize that they must consider the laws on sex discrimination in formulating their personnel policies.

Another legislative landmark, the Equal Pay Act of 1963, mandated equal pay for women and men doing similar work. On the local level a number of human relations commissions or fair employment commissions have been established. New York City, for example, has an effective Human Relations Commission to which problems of job discrimination are constantly being referred.

The field of drafting was a fully male dominated preserve until the early 1900s. There were some well-qualified women who found acceptance in the trades, although the first women pioneers in the drafting rooms of the late nineteenth century were indeed novelties. In 1870 there were only 13 women in the entire country working as drafters or designers. Of these, Harriet Irwin in 1869 pioneered as the first woman designer of dwellings on a professional basis. Louise B. Bethune began her apprenticeship as an architectural drafter in 1881 and went on to make a career as a designer of many distinctive structures. Similarly Sophia G. Hayden, beginning in 1891, earned a reputation as a capable designer. The two world wars found many women in drafting rooms where they helped design the plants, weapons, tanks, ships, planes, and the other materials of war which helped to bring victory.

In 1960 there were 15,750 women employed as drafters in the United States, amounting to 7 percent of the total number employed in the drafting profession. By 1991 this had increased to about 16 percent, or about 37,000 women in the field of drafting. It appears, therefore, that drafting offers a worthy career to qualified women. Most of the nation's high schools offer training in mechanical and technical drawing, thus making it possible for female secondary students to start in the trade. Nor is this opportunity confined to high school girls; women out of school who are seeking an occupation commensurate with their interests and background can get similar training in extension courses, community colleges, and other sources cited in this book.

One of these sources is the federally sponsored and state controlled system of apprenticeship described in another section of this manual. According to *Occupational Outlook Quarterly,* published by the United States Department of Labor:

> Women face many unique obstacles to apprenticeship—traditionally a male domain. Although more women are entering apprenticeship programs and being accepted by their male peers, many feel that they are breaking into a man's world—that they need much courage and self-confidence in addition to the abilities required of all apprentices. They face stereotyped attitudes held by many of their male coworkers. For example, men often try to protect women from heavy or dirty work, believing that women are too frail to handle it. On the other hand, some men make work even harder for women, because the men feel that the women don't belong in the trade. A study of apprenticeship programs in Wisconsin concluded: "... The barrier to women is not the difficult or dirty nature of some of the jobs, but the breaking of a taboo and the treading onto a territory that has remained the preserve of its male initiates." In the future, government initiatives and more women in apprenticeships will help to change attitudes with which women today must deal.

In some areas, women may actually experience certain advantages over their male counterparts in pursuing drafting careers. For example, employers will sometimes give preference to female job applicants in an effort to meet government guidelines or company policies designed to foster equity in hiring. They may also follow similar practices in awarding promotions of existing workers.

In addition, female students may qualify for special financial aid and support services. Many colleges and schools offer such programs supported through vocational education funds provided by the federal government. Special Gender Equity programs and Single Parent/Homemaker programs provide women with a variety of support services. These include career, academic, and personal counseling, and in some cases funds for everything from tuition and books to transportation and child-care expenses.

Increasing numbers of women have been entering fields related to drafting such as engineering. Professional schools in many technical fields are now enrolling more women, and their numbers will probably continue to grow.

THE ROLE OF THE DRAFTER TODAY

The drafter is vitally necessary in the increasingly complex technological civilization of today. Even though he or she forms but one link in the chain of production, it is the vital link that transforms *command* to *execution*. It would be impossible for a machine designer, for example, to transmit to a machinist the design for a special machine by word of mouth. Or can you imagine the erection of an Empire State Building by verbal instructions from the architects to all the people engaged for its construction? It must be done through the drafter's drawings and specifications. He or she, therefore, has a unique opportunity to observe and be part of the beginnings and final production of many inventions, discoveries, and ideas that are rapidly changing the world.

DRAFTING THROUGH THE AGES

There are three commonly used methods of conveying thoughts, concepts, and ideas among human beings, that is, of communicating with one another. Two of them are the spoken word and the written word, which are expressed in the innumerable languages and dialects of the peoples of the world. The third method is that of pictorial or graphic representation; that is, the use of pictures and drawings. All peoples, regardless of differences in languages, can understand pictures that are universal in appeal.

From the beginning, people used pictures to express themselves. Such pictures are still visible on the walls of caves in France, Italy, Spain, and other places. The primitive people who lived in these caves left behind vivid representations of animals (some long extinct), hunters, religious rituals, and the tools and weapons common to their time and place. Similar artistry has been found in many places on the earth, such as Mesopotamia. But as time went on and people took to agriculture, manufacture, and trade—settling into villages, towns, and cities—needs arose for the design and manufacture of more sophisticated tools, machines, structures, and devices. These demands ultimately led to the development of designers, builders, and others who needed and evolved a type of informative drawing which would communicate all the necessary directions and instructions for manufacture.

An ancient Sumerian statue now on display in Paris' Louvre Museum depicts Gudea, the ruler of ancient Lagash of 4,400 years ago, with the floor plan of a temple drawn on the clay tablet on his knee, an architectural drawing very similar to those being made now. There is in existence an Egyptian papyrus dated about 1500 B.C. which is in effect an architectural drawing showing the front and side elevations of what appears to be a religious shrine. There also exist Egyptian architectural plans engraved in stone. The ancient Egyptians were an advanced people in many respects, especially in applied mathematics and science. If we could by some miracle examine their drawings and calculations for the building of the pyramids, we would probably be greatly impressed by their grasp of fundamentals as well as fascinated by this glimpse of ancient methods and practices.

The writings of Roman scribes, such as Vitruvius, make mention of drawings used by the Greeks in erecting their beautiful temples, stadiums, and other structures. Phidias, their great architect and supervisor of the designing of the Parthenon, was also an artist and sculptor of high renown. It is inconceivable that he did not use a systematic type of drawing in his construction work. The Romans, the greatest civil engineers of their day, used drawings that showed a firm grasp of the fundamentals of orthographic or pictorial representation. However, it took Leonardo da Vinci, that universal genius of the Renaissance, to bring a high degree of sophistication to this type of drawing. His notebooks are full of exploded views, perspective sketches, section views, and the like, as well as the beginning of a mathematical basis for all this.

Following the example of Leonardo da Vinci, over two hundred years ago Denis Diderot, a famous French philosopher, began work on the great French encyclopedia of the trades, crafts, and industries of the period. For twenty-five years, assisted by the best drafters, artists, engravers, and printers of the period, he created thirteen of the largest sized folios, or books, filled with drawings and engravings illustrating all aspects of over 120 trades, crafts, and industries as practiced in France toward the end of the eighteenth century. Among these were metalworking, cannon founding, weaving, printing, fashion, cabinetmaking, tanning, and masonry, to name a few. These were shown in engravings and drawings in sequences of manufacture from raw materials to the finished article, so that one could visualize the process. Included were diagrammatic drawings such as cutaways of windmills, printing presses, wine presses, iron-forging machinery, and the like. There were perspective views, section views, exploded views, and other drawing methods familiar to present-day drafters. The encyclopedia embodies all that was at that time known of drafting, illustration, and engraving, and of printing from copper plates produced by artists and engravers. A two-volume excerpt from the encyclopedia was published in this country in 1975 by Dover Publications of New York City.

Gaspard Monge (1746–1818), a Frenchman, finally systematized the drawing methods of da Vinci, Diderot, and others into a descriptive geometry that applied rigid mathematical analysis to pictorial representation. Modern methods of mechanical drawing are based on Monge's system. His theories were treated for a while as a military secret by Napoleon Bonaparte.

These ideas were brought to the United States by Claude Crozet, another Frenchman, who taught them to the cadets at West Point in the 1820s. Thomas Jefferson, who lived in France for some time, undoubtedly used some of these theories in laying out the plans for Monticello, his famous home in Virginia.

The continuing Industrial Revolution of the nineteenth century created a great demand for mechanical drawings. All sorts of industrial machines, steam engines, boilers, power devices, tools, agricultural machinery, weaving and textile machines, and other mechanical devices were being invented and produced at a great rate. Buildings to house these machines also had to be built. These incorporated new concepts in architectural design, such as the use of steel. Mechanical and architectural drawings were often very elaborate, color was often used, and drafters took great pains in producing drawings that frequently resembled works of art.

The advent of iron and steel as commonly used materials for building construction revolutionized the profession of architecture and the techniques of architectural drafting. It also gave rise to new specializations, such as structural drafting. From antiquity to the end of the eighteenth century, the principal materials used in Western building construction were wood, stone, and brick. But because massive stone blocks had to be used for larger buildings, architects were limited in their designs, particularly as to height, and massive stone walls were necessary to carry the weight of floor and other interior arrangements. While iron had been known and used for centuries, its use was limited by an inadequate technology of production and by the fact that the iron so produced possessed no great strength and was easily affected by atmospheric conditions, which caused excessive rusting and decay. However, as the industrial revolution gained momentum, toward the end of the eighteenth century, a better grade of cast and wrought iron began to be produced and used in structures in the forms of posts, ribs, pillars, and the like.

With the success of the Bessemer process of making steel in large quantities and at acceptable costs, architects had at their command for the first time a comparatively inexpensive, light, and plentiful substitute for the heavy stone they had had to use before. In addition, this material could readily be rolled and formed into any sort of structural shape. This led to the first buildings erected entirely on steel structural forms; stone, brick, and glass formed the exterior facade but had no function in carrying any of the building load. This led to the invention of the American skyscraper. The first such building to be put up in the United States was designed by architect William LeBaron Jenney and was built in Chicago in 1885. This was followed by the designs of Louis Sullivan, who is often called the true inventor of the modern skyscraper and who also built in Chicago. It should be mentioned that making the tall buildings feasible and practical was the invention and development of the elevator and modern electrical technology.

With the advent of the automobile early in the twentieth century, the development of mass production, the experience of two great world wars, and the advances in aeronautics, electronics, and use of computers, the pace of industry and construction has quickened enormously. Automation is fast replacing much human effort and electronic data processing becomes ever more commonplace. Drafters have come into their own but no longer have time to make elaborate drawings. New processes, such as photography and the use of computers, have been developed to help them in their work, but the fundamentals remain the same. Modern drafters, although no longer required to be artists, must still be able to produce neat, accurate, and usable working drawings and must possess technical abilities in the sciences, mathematics, and technologies undreamed of by their predecessors. As for the future, drafting will become even more important as the vital link in the chain between idea and fulfillment. All over the world, breakthroughs in scientific research leading to new technological advances are sure to transform the world even further. These new technologies will require the drawings of the drafter for expression.

CAD-CAM: THE NEW TECHNOLOGIES IN DRAFTING

The impact of new technologies is being felt in the drafting room, too. To quote from the magazine *Product Engineering:*

> The drafting room is undergoing a revolution of sorts. Many of the tedious, time-consuming and routine aspects of drafting are being taken over by automatic equipment and as this shift takes place the day by day activities of engineering and drafting are changing.
>
> With this new equipment, design time is compressed to a great extent by the interaction common between designer, computer, and cathode ray tube (CRT) or digitizer. Commonly one can start with a rough drawing and within minutes get a display of a layout or drawing on a CRT; changes and modifications can be made at once, and a finished drawing achieved on the first go.

The technological changes in drafting are being implemented by the electronic computer, manually and automatically controlled drafting machines, cathode ray tube scanners and recorders, microfilm devices, and graphic man-machine consoles. In fact, according to *Occupational Outlook Quarterly,* "CAD may represent the greatest increase in productivity since electricity."

A quick look at today's newspaper want ads reveals the "new look" of modern drafting with ads such as the following:

> Drafter M/F—Experienced in Automatic Drafting—WE WANT YOU—excellent opportunity—phone to arrange interview. . . .
>
> Drafter, Computer M/F—Fast growing company offers opportunity for aggressive junior drafter—excellent opportunity to be trained in the fast growing field of computer graphics. . . .

The U.S. Department of Labor publishes a *Glossary of Technological Advances* that includes the following terms related to computer-assisted drafting:

Time-shared computer systems: Allows a person with a problem to communicate directly with the computer. Services more than one customer at a time and uses their free time to process other jobs.

Equipment to reproduce and store drawings: Accurately records the contents of drawings and stores them conveniently and compactly.

Numerically controlled plotters and drafting machines: Produces drawings using instructions and data from previously prepared tape or directly from the computer.

Cathode ray tube recorders: Interprets instructions and assembles the required picture on the face of the cathode ray tube. Then the information is photographed and printed.

Photo-composition device: An operator composes the drawing by mechanically selecting the proper symbols. When the symbols are properly positioned on the viewing screen, they are photographed.

Digitizing machine: A finished drawing is traced under manual control and the required coordinate positions are recorded and processed by a computer.

COMPUTER GRAPHICS

All these developments have given rise to an extension of drafting techniques known as computer graphics. Courses in this advanced electronic machine technology, along with standard drafting curricula, are now available in many community colleges, junior colleges, and technical institutes. It should be emphasized here that this technology is not going to make the drafter obsolete. It is intended as a sophisticated extension or tool that will allow an increase in the drafter's efficiency and work output.

THE TECHNOLOGIES OF THE FUTURE

We are now in that stage of computerized automation where it is possible to glimpse what manufacturing, construction, and other facets of our commercial and industrial systems will look like in the future. Computers are becoming so refined that it seems there is no limit to their possibilities. For example, more and more of the machine tools that cut, form, and otherwise produce the multitudinous plastic, metal, and wooden objects that form our cars, airplanes, trains, and the rest are now operated automatically by computerized numerical controls. In fact it is now possible to attach a computer to a series of machines, which are then directed through an entire manufacturing process.

In recent years, there has also been a marked increase in the design and use of robots, automatic devices that can be used in assembly lines or on machines to do a repetitive job. At automobile plants, combinations of robots and computerized production systems are carrying out entire processes—preparing the whole cylinder block, welding the entire frame, painting and finishing the body, and performing much of the final assembly.

All this had naturally led to the idea of letting computers run an entire factory. It is already being done in various industries, including chemical and oil processing plants, where entire processes are controlled by computers. The printing trades have gone through a revolution that has discarded most previous methods of hand composition. In their place has come a computerized operation by means of which the entire printed page can be produced automatically through computer terminals, phototypesetters, and other automated devices.

What all this means to the future of the drafter in the United States is quite evident. All the equipment, machines, buildings, conveyors, computers, tools, and the rest must be designed and detailed into working drawings before such equipment and plants can be built and put into production. Engineers, architects, designers, and drafters will be needed for this and other tasks mentioned in this book, and their employment will be assured well into the future.

However, despite all the marvels of electronic technology, it will still be necessary to train drafters in the basics and fundamentals of the trade. They will still function as a vital and necessary link between the dream, the design, and the fulfillment. In addition, advanced technology will place an emphasis on ever more technical education to improve drafters' skills and enable them to keep up with modern methods. Those without the advanced skills will probably find it difficult to hold on to anything more than a minor job in a drafting room.

THE OUTLOOK

Future prospects for employment in drafting appear promising. In its 1990–91 edition, the *Occupational Outlook Handbook* noted the following:

> Drafters held about 319,000 jobs in 1988. About one-third of all drafters worked in engineering and architectural services, firms that design construction projects or do other engineering work on a contract basis for organizations in other parts of the economy; and about one-third worked in durable goods manufacturing industries, such as machinery, electrical equipment, and fabricated metals. Drafters also were employed in the construction, transportation, communications, and utilities industries.
>
> Over 13,000 drafters worked in government in 1988, primarily at the State and local level. Most

> drafters in the Federal Government worked for the Department of Defense.

The *Handbook* went on to say:

> Employment of drafters is expected to grow about as fast as the average for all occupations through the year 2000. Industrial growth and the increasingly complex design problems associated with new products and processes will greatly increase the demand for drafting services. However, greater use of CAD equipment—which increases drafters' productivity—is expected to offset some of this growth in demand. Although some in the field had expected that CAD systems would decrease drafters' employment, this has not occurred in most situations were CAD systems have been installed. In fact, it now appears that productivity gains from CAD have been relatively modest. One reason is that CAD systems have been used to produce more variations of a design. As in other areas, the ease of obtaining computer-generated information stimulates a demand for more information. Although growth in employment will create many job openings, most job openings are expected to arise as drafters transfer to other occupations or leave the labor force.
>
> Drafters are highly concentrated in industries that are sensitive to cyclical swings in the economy, such as engineering and architectural services and durable goods manufacturing. During recessions, when fewer buildings are designed, drafters may be laid off.

The increasing demand for drafters is not of recent origin. Back in 1958, the Veterans' Administration published a report on the "Employment Outlook for Technicians" which stated:

> Technicians who work with engineers and scientists are among the *fastest growing occupational groups* in the United States. In recent years, the needs of the nation's defense program, added to those of the expanding civilian economy, have greatly intensified the demand not only for engineers and scientists but also for technical workers with less training than the technicians with whom this report is concerned. These technicians, whose jobs generally require a combination of basic scientific and mathematical knowledge and manual skill, participate in research and development work in designing, producing, and maintaining the machines and materials of our increasingly complex technology.

In this connection, it was estimated that of every ten technicians in industry about five are engineering and physical science aides, three are drafters, and the remainder are in other categories.

ECONOMIC SITUATION IN THE UNITED STATES

During the early and mid 1970s the United States experienced severe economic depression with corresponding increase in unemployment. Recovery gradually brought unemployment down to less than 6 percent, but a combination of economic factors with double-digit inflation, drastic rise in the cost of imported oil, and the decline of the automobile industry brought another economic depression in 1980 with resulting unemployment. The economy improved later in the decade, only to be followed by a period of recession in the early 1990s.

Despite this, drafters did not suffer as much as members of many other occupations. All types of drafting specialties continue to be in demand: mechanical, architectural, piping, construction, marine, electrical, electronic, tool design, technical illustrations, and others. In addition, despite increased unemployment in many areas, school systems across the country have experienced difficulties in getting teachers for such specialized subjects as mathematics, science, and particularly in vocational subjects, in which category drafting is included; this seems due to the high wages being paid by private industry.

Even with economic fluctuations, the impact on numbers of drafting jobs seems to have been minimal. As reflected in the figures cited previously, a steady if modest growth in demand for drafting personnel is expected over the next decade, regardless of routine economic ups and downs.

The economic setback of the early 1970s brought with it a totally unexpected situation. Beginning with an oil embargo imposed upon the United States by the oil-producing nations of the Near East, an energy crisis developed that threatened drastic realignments in the economic life of our country. However, all this did lead to some positive developments. The nation rallied to contain the situation by increasing exploration and creating new technologies and industries concerned with generating power without importing oil. The last two decades have seen continued research and development in innovative technologies such as the further development of atomic energy, the extraction of oil from shale rock, the use of the sun's energy, the design of new engines and prime movers, and other techniques. Other options being explored have included enhanced oil recovery, unconventional gas recovery, industrial fluidized-bed coal combustion, low head hydroelectric power, passive solar energy, solar hot water, and industrial-process heating. All these efforts in research and development have helped create new and diverse employment opportunities for persons interested in drafting careers.

The following newspaper ads are representative of needs for drafting personnel:

> Designers—Marine—Hull, Piping, Machinery, mechanical, electrical, HVAC. . . .
>
> Designers, Engineers, Drafters—100's of openings—Power plants, chemical, Petro Chem; Electrical, Piping. . . .
>
> Drafters; Structural, U.S. experience. . . .
>
> Drafters; Food Services and Equipment Planning. . . .
>
> Draftsmen/Women—Architecture, interior store planning. . . .
>
> Mechanical Draftsmen/Women—experience in heavy equipment. . . .
>
> Tool and Die Designer—supervise die-makers. . . .
>
> Drafting Trainee—knowledge Mechanical Drawing—learn layouts for Die Design. . . .
>
> Drafter/Illustrator—isometric projections, exploded views, schematics.

Ads such as these can be seen in practically all the metropolitan papers of our country. Drafters, sure of their skills and their important place in the economic life of our nation, can find ample opportunities to practice and further their skills.

PERSONAL ATTRIBUTES NECESSARY FOR SUCCESS IN FIELD

A man who spent a career in drafting and related fields tells of working once as a checker in the drawing and design department of a large concern. It was his particular task to examine every completed design and its details for correctness, for possible conflicts, and for errors that might have crept into the dimensions and specifications. In this particular drawing room, the drawings of one drafter were the cause of the greatest amount of trouble. In fact, after a while all his drawings became suspect. Although he was a hard worker, he seemed to be careless and his work had to be watched with utmost care. Despite all precautions, however, errors in his drawings sometimes did escape all checking and went into the shop to cause endless trouble and, at times, great expense. Finally, the particular drafter had to be discharged as being too much of a risk, despite his ability.

To illustrate the necessity of accuracy in a drafter's work and her or his need for a thorough background in the technical aspects of a trade, consider the classic story of one who had laid out and detailed the design of a huge cast iron cylinder to be used in a steam generating plant. The vessel was too large to be cast by the usual methods and, therefore, had to be molded directly into the sand on the foundry floor, entailing considerable labor. Meanwhile, the drafter had been instructed to figure out the amount of iron that needed to be melted in the foundry cupola to fill the mold. When all was ready, pouring of the molten metal began. Long before completion of this process, it became evident that not enough iron had been melted to completely fill the mold. Of course, before another melt of metal could be provided, the iron already in the mold would have cooled and fusion would have been impossible. Consequently the operation was completely ruined. Needless to say, the drafter had meanwhile packed up his instruments and quietly faded away.

From this it can be seen that drafting is usually an exacting, confining, and sedentary occupation. The drafter, although sometimes in the field or shop for special work, is usually confined to an area containing a drawing board, a stool, and perhaps a reference table or bench or in the case of computerized operations, a computer system. Here he or she spends long periods of time concentrating on the task at hand, which does not permit much fraternization with neighbors or moving around. Not much physical energy is used outside of moving drawing instruments around or getting up once in a while to consult a handbook or some other source of technical information. The drafter generally works in clean, pleasant, well-lit surroundings. But the strain on eyesight is considerable, and bending over the board may, in time, cause stomach distress.

In planning a career in drafting, therefore, there are several essentials to reckon with. A person must be able to concentrate on a task for relatively long periods of time. He or she should be able to withstand confinement to one spot (as in most office occupations). He or she should not require the stimulus of constant give and take with other people or the need to move around and mingle with them.

The prospective drafter must have a liking for mathematics and science, some ability to visualize objects in full from pictures in the flat, an ability to solve problems, and some creative instincts that find satisfaction in the drawings produced. With modern drafting methods, the ability to work with computers may also be needed. One who is planning to be an architectural drafter should have some artistic flair, in addition. A careless and inaccurate individual will be very unhappy on the drawing board, where accuracy is most essential. Sloppy and careless work means great trouble and unnecessary manufacturing costs, and the creator of these undesirable conditions does not last long on the job.

The General Motors Corporation published a pamphlet called *Drafting—Can I Get the Job?,* from which the following extract was taken. It can serve as a useful guide to one thinking about drafting as a career and desiring some guidance on the subject.

If you are considering drafting and designing as a career, and are wondering whether or not you are suited for this kind of work, here are a few indicators that might help you decide.

- Do you like to draw? More particularly, do you like to make the kinds of drawings from which something can be built? If you do, you've met the first requirement.
- Can you visualize objects; that is, can you see in your mind how an object looks, or might look, before it is drawn on paper? This ability is one which all draftsmen must possess.
- Are you mechanically inclined? Since drafting requires a good knowledge of mechanisms and their operation, an aptitude for mechanical things is a real asset to prospective draftsmen. Perhaps you've already had some sort of me-

chanical aptitude test in school and know how you score in this area. If not, perhaps you could arrange to take such a test soon.

- Do you have access to a home workshop? Do you like to make or fix things around home?
- Are you neat, systematic and thorough in your school work and outside activities?
- Do you like to work as part of a team?
- Would you prefer to work in one place, rather than move around constantly, like a salesman, for example?
- Do you like to work on projects which involve considerable detail?
- Are you interested in science and mathematics?

These are some of the interests and qualities which most experienced draftsmen possess, and if you can claim them now, drafting may be the career for you.

Drafting can be a most interesting and rewarding occupation. Most drafters can recall days at the board or computer when, lost in some problem, the time went by unnoticed as the difficulty was being solved in the lines and curves of a design or layout that later went into the shops. It was quite a satisfaction, after the inevitable "bugs" had been eliminated, to see the finished article work and to know one had contributed to its creation. Often the drafter has a part in the beginning of many a famous structure, device, machine, or process. Somewhere, sometime, some drafter laid out on a drafting board the beginnings of the atom bomb, the first missile, the Polaris submarine, and the space capsule that carried astronauts to the moon. Somewhere, someplace, a drafter is now putting on paper the final designs and details for amazing new space vehicles and their accessories. Wherever advances in science and technology are being made, the drafter is in the vanguard helping to translate ideas into realities.

GRADES OF DRAFTERS

Drafters are graded or classified in accordance with experience and ability. No one grading system is prevalent in the United States; however, there are several that seem to be more in use than others. For example, traditional drafters sometimes began as *tracers,* or copiers, skilled in making ink tracings or copies of drawings for permanent record. There is not too much demand for this type of work in our fast-changing technologies. Most drafters with some training at school or elsewhere will start as *junior drafters* or *detailers,* who are entrusted with making working drawings of details or parts from an assembly drawing, from sketches, or from other sources of information. The next grade is that of *senior drafter,* who has the beginnings of a designer and can lay out or draw assemblies, design, and details from notes, sketches, and instructions. Top grade is that of *leading drafter* or squad or section leader, who is fully qualified as a designer, often working with a group of detailers.

Another system of job grading starts with a *learner detailer,* who is a beginner just out of school with very little or no actual work experience. He or she knows some of the techniques of drawing and generally is used as a tracer and detailer of simple parts. The next step is a *detailer,* who with experience is able to handle most detail work and can be entrusted with some responsibility. He or she will graduate to a *senior detailer* and by then can make design layouts on which design and problems of details can be solved accurately. Such activities may involve machinery, tools, installation of machinery, materials, plant layouts, and/or other details of design. The next step would be that of a *junior designer,* who can be entrusted with some original design work under the direction of the architect, engineer or technical manager. From this he or she can advance to *layout drafter,* who can lay out or draw the most intricate of devices or products to accurate scale with all parts in proper relationship, working from any type of information. The top grade is that of *designer.* By this time the drafter has an intimate knowledge of mechanisms, devices, standards, machinery, metals, products, and so on and can be given the most complex design problems to solve. He or she works directly with architects, builders, engineers, scientists, technicians, and inventors. In large drafting offices, a top drafter is often used as a *checker,* a responsible job involving inspection of all finished drawings for errors of design, detail, or other factors affecting the finished product and returning them for correction if necessary.

In addition to job-level designations, many drafters' positions are defined by their area of specialty. For example, the U.S. Department of Labor lists these specialized drafting positions in the *Occupational Outlook Handbook:*

> *Architectural drafters* draw architectural and structural features of buildings and other structures. They may specialize by the type of structure, such as schools or office buildings, or by material, such as reinforced concrete or stonework.
>
> *Aeronautical drafters* draw wiring and layout diagrams used by workers who erect, install, and repair

electrical equipment and wiring in powerplants, electrical distribution systems, and buildings.

Electronic drafters draw wiring diagrams, schematics, and layout drawings used in the manufacture, installation, and repair of electronic equipment.

Civil drafters prepare drawings and topographical and relief maps used in civil engineering projects such as highways, bridges, flood control projects, and water and sewage systems.

Mechanical drafters draw detailed working diagrams of machinery and mechanical devices, including dimensions, fastening methods, and other engineering information.

CHAPTER 2

EDUCATIONAL PREPARATION

If you want to follow a drafting career, the right educational preparation is a must. As with many fields, no single level of educational preparation stands as a universal requirement. But a high school diploma is usually the minimum. In some cases postsecondary training is needed, either as entry-level preparation or as continuing education after employment begins. This generally consists of a certificate program taking a year or less to complete, or an associate degree requiring two years of full-time study.

Whatever level of education you pursue, the most important factor is to acquire a good foundation in technical education. This means more than simply taking classes in technical drawing. You must also have a solid understanding of science and mathematics, for the drafter is often called upon to make mathematical calculations and to apply basic scientific principles.

For those planning to pursue a drafting career, a typical high school curriculum would be as follows:

English	4 years
Social Studie	3 years
Mathematics (Algebra, Geometry, Trigonometry)	3 years

Science (General Science, Physics, Chemistry)	3 years
Technical Drawing	3 years
Health Education, Electives, Foreign Language	

In a typical vocational-technical high school, the student enrolled in a typical technical curriculum is in shop, laboratory, and class for a total of forty periods or thirty hours per week. Of this, the student spends twenty periods or fifteen hours a week in a technical specialty in laboratory or shop (above the ninth year). In addition, the student has classes in mathematics and in the sciences pertaining to his or her specialty. Besides this, the usual high school program of English, social studies, and basic science must be completed. In this way the technical student receives a standard high school education in addition to specialized technical subjects, for it is vitally necessary that we have an informed and enlightened citizenry as well as qualified technicians.

STAY IN SCHOOL

All this is a bare minimum, but it is important here to emphasize that for an aspiring drafter it is an absolute necessity. Without a high school diploma the obstacles to a career in technical work are almost insurmountable. If you have ideas of entering the field and are tempted to leave school for a job of some kind that gives the promise of some immediate cash, don't give way or you will surely regret it. Even if you do not finally go into drafting, your high school diploma is a basic requirement for any decent job in the 1990s or the twenty-first century.

AFTER HIGH SCHOOL

Vocational-technical high school courses are a good starting point for the aspiring drafter. After high school, however, he or she will probably need further technical education. Depending upon the specialized area in which the drafter is interested, further training in design, mathematics, science, and specialized technologies will be needed for advancement. These courses can be found in many community, junior, and technical colleges, as well as in other postsecondary institutions.

To help the beginning drafter plan a future, and as a guide to those with some experience, who would benefit from additional education, typical courses in drafting and other technologies as offered in two-year colleges are presented in the following list. These schools usually provide a rounded curriculum which, in addition to these specific courses, also provides for mathematics, science, social studies, and electives in sufficient depth to lead to a two-year associate degree. Many four-year colleges will accept these associate degrees for credit to permit a student to get a bachelor's degree within another two years. However, if the drafter cannot devote the necessary time for a degree course he or she usually can take just the technology courses in the evenings for technical advancement.

Drafting

Building Construction Drafting
Advanced Building Construction
Engineering Drawing
Industrial Drafting
Computer Aided Drafting I
Computer Aided Drafting II
Site and Survey Drafting

Mechanical Design Drafting
Advanced Mechanical Design
Technical Drawing
Symbols and Schematics
Blueprint Reading
Structural Drafting

Construction Technology
Projection Drawing
Construction Techniques
Elements of Construction
Small House Architecture
Construction Methods and Practices
Statics and Strength of Materials
Computer Programming

Mechanical Technology
Industrial Processes I
Engineering Drawing I
Industrial Organization
Manufacturing Processes
Industrial Processes II
Engineering Drawing II
Mechanics and Strength of Materials I
Computer Programming

Construction Technology
Industrial Architecture
Structural Steel Fabrication
Elements of Structural Steel Detailing
Elementary Surveying
Structural Steel Design
Construction Estimating
Structural Steel Detailing
Route Surveying
Reinforced Concrete Design

Mechanical Technology
Production Processes
Basic Thermodynamics
Machine Design I

Mechanics and Strength of Materials II
Metallurgy I
Computer Courses

To indicate the sort of technical knowledge the drafter needs to get along in her or his work, catalog descriptions of several of these courses are given as follows.

TECHNICAL DRAFTING I–II

Introduces technical drafting from the fundamentals through advanced drafting practices. Teaches lettering, metric construction, technical sketching, orthographic projection, sections, intersections, development, fasteners, theory and applications of dimensioning, and tolerances. Includes pictorial drawing, and preparation of working and detailed drawings.

GEOMETRIC TOLERANCING

Teaches use of a positional tolerance system, its relationship to coordinate tolerance systems, and other aspects of U.S. standard drafting practices.

INTRODUCTION TO ELECTRICAL/ ELECTRONICS DRAFTING

Teaches applications of drafting procedures with emphasis on working and functional drawings and direct applications to electrical and electronic components and circuits.

ELECTRICAL AND ELECTRONIC DRAFTING I–II

Teaches the design of block and logic, schematic and wiring diagrams, house wiring plans, printed circuit boards, and card cages.

ELECTRICAL/ELECTRONICS BLUEPRINT READING

Presents an interpretation of basic shop drawings, conventional symbols, terminology, and principles used by the mechanical draftsman. Explains common electrical and electronic symbols, wiring diagrams, schematic drawing, and application of wiring diagrams, schematic drawings, and application of wiring diagrams.

DESCRIPTIVE GEOMETRY FOR DRAFTING
Gives analysis and graphic presentation of the space relationship of fundamental geometric elements as point, line, plane, curved surfaces, development and vectors.

MACHINE BLUEPRINT READING
Introduces interpreting of various blueprints and working drawings. Applies basic principles and techniques such as visualization of an object, orthographic projection, technical sketching, and drafting terminology. Requires outside preparation.

ARCHITECTURAL BLUEPRINT READING
Emphasizes reading, understanding, and interpreting standard types of architectural drawings including plans, elevation, sections, and details.

SURVEY OF COMPUTER-AIDED DRAFTING
Surveys computer-aided drafting equipment and concepts. Develops general understanding of components, operations, and use of a typical CAD system.

COMPUTER-AIDED DRAFTING AND DESIGN I
Teaches computer-aided drafting concepts and equipment (Autocad) designed to develop a general understanding of components of a typical CAD system and its operation.

COMPUTER-AIDED DRAFTING AND DESIGN II
Teaches working drawings and advanced operations in computer-aided drafting.

MACHINE DESIGN
Teaches elements of design and development of cams, gears, charts, graphs. Explains true position of dimensioning and military standards.

JIG AND FIXTURE DESIGN
Teaches design of jigs and fixtures through research and a study of shop processes and materials.

FINANCIAL AID FOR THE POST-HIGH SCHOOL STUDENT

Many graduates or potential high school graduates who have taken courses in technical drawing and who would like to continue in a two-year technical curriculum at a junior college, community college, or technical institute often find it difficult to do so because of financial problems. For them there exist numerous scholarships, student financial grants, student loan plans, and other forms of help. Many of these are of national scope, such as the federal Pell Grant; others are restricted to certain localities and even neighborhoods. Still others are meant for members of a certain group, such as children of war veterans or of union members. There are also various student aid plans offered by State governments and a wide range of private organizations.

The student's first step should be to consult the guidance counselor in her or his school. If out of school and unable to reach a former counselor, he or she can obtain guidance help at the local office of the Federal Employment Service, at the local Manpower Human Resources center, or at the nearest office of the Veterans' Administration. With the help of a counselor or any of the above agencies, the student can determine the institution or institutions to which to apply for entrance. He or she can then write to them for admission application blanks and applications for financial aid. When the documents arrive, they should be filled out with the help of a counselor and parents or guardian. All questions should be answered fully, holding nothing back about the student's financial status or that of the family. If the applications are successful, the student can then proceed with her or his post-high school education.

Available federal aid programs include the following:

- Pell Grants (awards that need not be repaid)
- Supplementary Educational Opportunity Grants (similar to Pell Grants)
- College Work-Study Awards (part-time jobs that pay at least the federal minimum wage)
- Perkins Loans (low-interest loans)
- Stafford Loans (low-interest loans)
- Plus Loans (loans for parents who want to borrow for their child's educational costs)
- Supplemental Loans for Students (loans made directly to students)

For information about these programs, call the Federal Student Aid Information Center at 1-800-433-3243 or write to P.O. Box 84, Washington, DC 20044.

If for some reason a student cannot receive federal or state aid, there are other resources. If either of the student's parents belongs to a union, for example, there may be some type of scholarship or award offered by that organization. Again, if a member of the family belongs to a fraternal organization, it too may offer some form of aid for which the student can apply. Other contacts might come from the family's religious center; or if there are war veterans in the family, he or she can inquire of the American Legion and other veterans' organizations. He or she can find out if there is a student aid agency in the state and can get in touch with it, or the family's bank can be approached for a loan under the Federal Student Loan Plan. If the student is persistent and determined, he or she will finally find the necessary help.

As an aid in the search for financial assistance the following reference books can be most useful. They can be found in libraries and bookstores.

- *Directory of Financial Aids for Women.* Reference Service Press, 1991.
- *Free Money for College.* Facts on File, 1990.
- *How to Go to College Free.* Probus Publishing, 1991.
- *Keys to Financing a College Education.* Barron, 1990.
- *Lovejoy's Guide to Financial Aid.* Arco, 1989.
- *The Scholarship Book: The Complete Guide to Private-Sector Scholarships, Grants and Loans for Undergraduates.* Prentice-Hall, 1990.

ON-THE-JOB TRAINING AND APPRENTICESHIPS

There was a time when it was possible to learn the drafting trade on the job. One could get a job as a "blueprint boy" in some drafting office and, over the years, pick up the manipulative skills and enough technology to become a drafter. But this haphazard approach is disappearing and, in fact, is barely possible in this day of quickly changing and expanding technologies. It is necessary to go through a planned, systematic learning process such as is provided by a high school course.

There are, however, other planned methods, such as an apprenticeship, which do include an on-the-job approach. Apprenticeship has come down to us in a modified form from the system used in the Middle Ages in Europe. At that time, it was the only road by which a youngster could acquire the skills required by a craft. At an early age he would be bound or indentured by contract to a master craftsman in whose establishment he would live for years until he reached manhood. During this time he was instructed in the particular craft involved. In return, he did his master's bidding, received no cash wages, and was practically the master's serf. At the end of this period, the apprentice took a difficult test in his trade. If he was qualified, he was then on his own, free to practice his craft for wages as a journeyman. In time, he too would become a master and instruct apprentices.

The English who settled on the eastern seaboard of America brought with them this system of apprenticeship. It became an accepted pattern for the mastery of a trade in this country. The method, with some variations, lasted well into the nineteenth century, and there are people still alive who learned a trade through such an indentured apprenticeship. One famous American who profited from the system was Benjamin Franklin, who was apprenticed by his father to his elder brother James. James was paid ten pounds to teach his younger brother "the mysteries" of printing, a trade that enabled Benjamin to make a start on a notable career as one of the foremost Americans of his day. Paul Revere, another man who is well known to history, learned the trade of silversmithing as an apprentice and then went on to become one of the greatest silversmiths in the colonies. His products are highly prized today and are considered museum pieces and collectors' items.

The lengthy and restrictive procedure of apprenticeship as it was once practiced is no longer possible or necessary in our day. But it is currently being practiced in a modified form under a federal statute. The Bureau of Apprenticeship and Training, a division of the U.S. Department of Labor in Washington, DC, encourages the formation of apprenticeship agencies in the various states of the union. These, in turn, encourage the formation of joint apprenticeship committees in various unionized trades. These committees operate under the auspices of labor and management. Beginners entering these trades become apprentices under the terms of an agreement which guarantees that over a certain number of years they will be taught their craft on the job in a systematic fashion and will also be provided after work hours with related instruction, such as mathematics, science, and blueprint reading, by the local board of education.

The Bureau of Apprenticeship and Training of the U.S. Department of Labor defines apprenticeable occupations as follows:

Apprenticeable Occupations

An occupation recognized as apprenticeable by the Bureau of Apprenticeship and Training is one which:

1. Is customarily learned in a practical way though a structured, systematic program of supervised on-the-job training.
2. Is clearly identified and commonly recognized throughout an industry.
3. Involves manual, mechanical, or technical skills and knowledge that require a minimum of 2,000 hours of on-the-job work experience.

4. Requires related instruction to supplement the on-the-job training. Such instruction may be given in a classroom, through correspondence courses, self-study, or other means of approved instruction.

Most apprentice programs gill admit candidates between the ages of 18 and 28. In addition, the requirements for admission usually cover education, aptitude, and physical condition. Education is important, a high school diploma usually being a prime consideration. Many trades place emphasis on ability to read and write adequately and on courses in mathematics, science, and mechanical drawing. As for aptitude, many sponsors look for people who have the mechanical and mental abilities to master the techniques and technology of a trade. To this end, most sponsors administer some form of objective testing.

A listing of the field and regional offices of the Bureau of Apprenticeship and Training can be found in Appendix D. These offices can be found in all sections of the country, and you can consult the one nearest to you for further information. In addition, you will find a listing of state apprenticeship agencies and their locations. You can also write to the Bureau of Apprenticeship and Training for a copy of its bulletin entitled "The National Apprenticeship Program."

LABOR-MANAGEMENT APPRENTICE AGREEMENT

Typical of this kind of arrangement in the drafting fields are the Apprenticeship and Training Standards developed by the American Federation of Professional and Technical Engineers, AFL-CIO, in cooperation with U.S. Bureau of

Apprenticeship and Training. They are used by locals of this union as guides in establishing local programs of apprenticeship in the drafting trades. Points that are covered in these standards include qualifications of candidates, credit for previous experience and education, term of service, agreement of indenture, supervision, probationary period, hours of work, salaries, periodic examination, ratio of apprentices to other employees, and so on. Of particular interest is the work schedule of the drafter apprentice, which is set forth as follows:

Schedule	*Approximate Months*
Tracing	3 months
Routine Department Work	3 months
Detailing	12 months
Engineering Changes	12 months
Simple Assembly Drawings	12 months
Subassembly Drawings	12 months
	54 months

In addition, it is recommended that the apprentice receive related instruction after work hours in machine shop practice, theory of drafting, design, intermediate algebra, technical illustration, advanced algebra, descriptive geometry, trigonometry, analytic geometry, elementary statics, and elementary strength of materials.

Work schedules for apprentices that formerly set rather rigid time allotments for each learning block or segment were modified in many instances by a plan developed and advocated by the Association for Manufacturing Technology. The plan stressed merit, not just time-served. Under this scheme, the apprentice advances at her or his own speed from one learning block to another through her or his own efforts. If, because of previous experience at school or elsewhere, or because of superior abilities, an apprentice can finish the work specified in a particular segment before the time allotted, he or she can go on to the next level. In this way it is possible for a beginner to finish an apprenticeship before the specified time and go on to journeyman status and wages.

The work schedule just described is designed for the apprentice in manufacturing or mechanical drafting. For the architectural drafter, the following schedule is usually recommended:

First Year

1. Filing of prints, tracings, and drawings.
2. Tracing of drawings, following special instructions as to methods and detail. Simple architectural details.
3. Making minor alterations to drawings or tracings, transcribing of same, and following instructions as to views, projection, and sections.
4. Drawing of simple construction details. Visits to jobs to learn application.

Second Year

5. Design of simple architectural motifs. Advanced structural details. Advanced materials study and their uses and limitations.
6. Developing preliminary drawings and relating suggested specifications to same for consultation purposes. Job visits to observe application of principles.

Third Year

7. Developing working drawings, integrating known specifications into them; minor job supervision.

Fourth Year

8. Developing working drawings, including rough layout of mechanical trades involved; designing of structures; supervision of jobs in

progress and preparing reports on same. Preparing prebid cost estimates.

In addition to this, the following related instruction is recommended: theory of orthographic projection, freehand sketching, architectural lettering, art-life drawing, layout architectural drawing in pencil and ink, basic Greek and Roman architecture, elementary engineering mathematics, technical illustration, trigonometry, strength of materials, estimating, basic building codes, computerized drafting, and other relevant subjects.

Similar programs designed to suit specialized areas such as marine drafting are also recommended.

Any aspiring drafter who successfully completes the recommended schedule of work experience and supplementary technical education is thereby assured of a firm foundation in the trade. On this foundation a good and satisfying professional career can be built.

For sources of further information, consult the field and regional offices of the Bureau of Apprenticeship and Training, listed in Appendix D, as well as the state apprenticeship agency offices listed.

EVENING COURSES

It is possible to gain basic training in drafting techniques through the many courses operated by public school systems throughout our nation, as well as in such post-high school institutions as technical colleges, community colleges, and schools of engineering and architecture. Practically every sizable town or city offers such an opportunity. Many such courses are offered in the evenings. Therefore, it is possible to work at your job and go to school in the evenings for the training needed to embark on a drafting career. For information, contact your local board of education's central office or local two-year colleges.

In addition, the list of junior colleges, community colleges, and technical institutes found in Appendix E can serve as an aid in getting detailed information about evening and other courses in drafting.

PRIVATE SCHOOLS AND CORRESPONDENCE COURSES

Many reputable vocational and private correspondence schools in this country offer courses in the basic elements of drafting, as well as in specialized and technical aspects of the trade. Such schools may be of interest to persons who cannot get the training in other ways and to veterans who are entitled to special educational benefits. The person contemplating enrolling in a private school should exercise a great deal of caution in selecting a school or a correspondence course, for government investigations have uncovered a few unscrupulous operators of such schools.

In selecting a drafting course, certain precautions should be taken. You should be cautious about the school that makes extravagant claims about its instructional program and job placement record. No school, no matter how efficient, can train a top-grade member of a trade in an unreasonably short time, nor can the graduate of such a course expect to receive top wages on a first job. The record of the school in securing jobs for its graduates should be checked if possible. It seems that all too often graduates in some courses cannot find the jobs that were promised. Make sure that you can get some or all of your money back if you are dissatisfied with the course. Be careful about signing a contract; if you can't interpret the small print, find someone who can. If a salesperson calls, don't allow her or him to pressure you into signing the contract at once. Take your time and check it as thoroughly as you can. Of course you will find most schools are reputable and will do all they can to help you, but it pays to be careful.

There is help available for finding a good private study course. Write to the National Home Study Council, 1601 18th Street, NW, Washington, DC 20009, for advice on home correspondence courses. Ask for its directory of accredited private home study schools. For help with private vocational schools, get in touch with the National Association of Trade and Technical Schools, 2251 Wisconsin Ave., NW, Washington, DC 20007.

THE JOB TRAINING PARTNERSHIP ACT

In 1962 President Kennedy signed into law a Manpower Development and Training Act. The act was designed for the retraining and consequent employment of persons who had lost their jobs because of the advance of technology and automation; for young persons who needed training in some skill in order to get a job; for out-of-school, jobless youths; and in general for those designated as underprivileged. The basic purpose of the act was to help people acquire skills for job placement purposes. Trainees fell into certain categories, including unemployed workers whose skills had become obsolete because of technological changes or because of industry relocation and shifts in job market demands. Another eligible group consisted of trainees who needed additional skills in their particular trades to become fully productive and employable. Still another category included young people sixteen years of age or over who were out of school and required skills training to qualify for a job.

Since 1962 the Manpower Act has been amended several times. In 1973 it was replaced by the Comprehensive Employment Training Act, popularly known as CETA. Then again in 1983, the legislation was amended under the Job Training Partnership Act (JTPA), which is still in effect.

To the prospective student, these federally funded programs represent an opportunity—if your family has a limited income or you meet other special requirements—to receive job counseling and even free training in a number of fields, including drafting. As a participant, you can have tuition paid, books purchased, and other key support in receiving the training needed to land a job. To get further information, contact the local board of education or the local regional office of the federal or state employment service. Large cities usually have neighborhood manpower offices. Another information source would be your state's department of education and its guidance services. It certainly is worth your while to make an effort to receive this training if you are eligible.

VETERANS' BENEFITS

After World War II, the United States embarked on a great experiment in education by enacting in 1945 a Veterans Readjustment Act, popularly known as the GI Bill of Rights. Its monetary benefits were the means of providing many thousands of ex-service persons with the opportunity for an education, ranging from the completion of elementary and high school courses, to trade and commercial training, up and through undergraduate and graduate school and professional degrees. Of course, this act expired some years ago, but the experience led the nation to enact similar bills for the benefit of veterans of the Korean conflict of 1952–55 and of other wars since that time, as well as for those who serve in times of peace.

The following is extracted from the Veterans' Administration's Bulletin #22-90-2, entitled "Summary of Educational Benefits Under the Montgomery GI Bill—Active Duty Educational Assistance Program."

ARE YOU ELIGIBLE?

You may be eligible for education benefits as a veteran or serviceperson if your active duty is under one of the following categories.

Category 1 - You enter active duty for the first time after June 30, 1985, and serve continuously for 3 years.

However, only 2 years of active duty are required if:

- You are now on active duty;
- You first enlist for 2 years; or
- You have an obligation to serve 4 years in the Selected Reserve (the 2 by 4 program). You must enter the Selected Reserve within 1 year of your release from active duty.

In Category 1, you must have your military pay reduced by $100 a month for the first 12 months of active duty. If you elect not to participate in this program, you may not change this decision at a later date.

You may also be eligible if:

- You were on active duty between December 1, 1988, and June 30, 1989.
- You withdrew your election not to participate.
- You had your military pay reduced by $100 a month.
- You completed the period of active duty you were obligated to serve on December 1, 1988. If you did not complete the period, your discharge must be for one of the reasons in the next paragraph.

You must obtain a high school diploma or an equivalency certificate before your first period of active duty ends. Completing 12 credit hours toward a college degree meets the requirement.

Category 2 - You had remaining entitlement under the Vietnam Era GI Bill on December 31, 1989, and served on active duty from October 19, 1984, to:

- June 30, 1988; or
- June 30, 1987, and you serve 4 years in the Selected Reserve after release from active duty. You must have entered the Selected Reserve within one year of your release from active duty.

You must have obtained a high school diploma or an equivalency certificate before December 31, 1989. Completing 12 credit hours toward college degree meets the requirement.

DISCHARGES AND SEPARATIONS

Your discharge must be "honorable." Discharges "under honorable conditions" and "general" discharges do *not* establish eligibility for education benefits.

A discharge or release for one of the following reasons could result in a reduction of the required length of active duty.

- Convenience of the government;
- Disability;
- Hardship;
- A medical condition existing before service; or
- Certain reductions in force (RIF);

OTHER ISSUES

You are not eligible if you graduated from a service academy and received a commission after December 31, 1976.

You are not eligible if you received a commission after completing a Reserve Officers' Training Corps scholarship program after December 31, 1976. This scholarship pays a stipend *and* all educational expenses, i.e., tuition, fees, books, and supplies. If you completed Reserve Officers' Training Corps *without a full scholarship,* you are eligible to participate in Chapter 30.

The following types of active duty do *not* establish eligibility:

- Time assigned by the military to a civilian institution for the same course provided to civilians;
- Time served as a cadet or midshipman at a service academy; or
- Time spent on active duty for training in the National Guard or Reserve. (NOTE: Active duty for training does count towards the 4 years in the Selected Reserve under the 2 by 4 program.)

HOW MANY MONTHS OF BENEFITS CAN YOU GET?

Generally, you have 36 months of entitlement under this program. However, if you did not complete your enlistment period, you earn only 1 month of entitlement for each month of active duty after June 30, 1985.

You may also earn 1 month of entitlement for each 4 months in the Selected Reserve after June 30, 1985. You can earn at most 36 months under this program.

You may receive a maximum of 48 months of benefits under more than one VA education program. If you used 30 months of Chapter 35, and are eligible for Chapter 30, you could have 18 more months. If you used 27 months under Chapter 34 before December 31, 1989, you could have 21 more months.

HOW IS ENTITLEMENT CHARGED?

You are charged one full day of entitlement for each day of full time benefits paid.

For correspondence and flight training, you use 1 month each time VA pays 1 month of benefits. If your full rate is $300, and you receive $900 for a correspondence course, you use 3 months. If your full time rate is $250, and you receive $3,000 for flight training, you use 12 months.

If you pursue a cooperative program, you use 80% of a month for each month of benefits paid.

For apprenticeship and job training programs, the charge changes every 6 months. During the first 6 months, the charge is 75% of full time. For the second

6 months, the charge is 55% of full time. For the rest of the program, the charge is 35% of full time.

VA can extend entitlement to the end of a term, quarter, or semester, if the ending date of your entitlement falls within a term, quarter, or semester. If the school does not operate on a term, quarter, or semester basis, entitlement can be extended for 16 weeks.

HOW MUCH EDUCATIONAL ASSISTANCE WILL YOU GET?

BASIC EDUCATIONAL ASSISTANCE

If you first entered active duty after June 30, 1985, and your first service obligation was more than 2 years, you will receive the following monthly rates:

Rates for Schooling

Monthly Rate	*Training Time*
$300	Full
$225	Three-Quarter
$150	One-Half

Rates for Apprenticeship and Job Training

Monthly Rate	*Period of Training*
$225	First 6 Months
$165	Second 6 Months
$105	Remainder of Program

If you first entered active duty after June 30, 1985, and your first service obligation was 2 years plus 4 years in the Selected Reserve, you will also receive the rates shown above.

If you first entered active duty after June 30, 1985, and your first service obligation was 2 years, you will receive the following monthly rates:

Rates for Schooling

Monthly Rate	*Training Time*
$250	Full
$187.50	Three-Quarter
$125	One-Half

Rates for Apprenticeship and Job Training

Monthly Rate	*Period of Training*
$187.50	First 6 Months
$137.50	Second 6 Months
$87.50	Remainder of Program

RATES FOR OTHER TYPES OF TRAINING

If you take cooperative training, you receive 80% of the full time rate.

If you take a correspondence course, you receive 55% of the approved charge for the course.

If you take flight training, you receive 60% of the approved charges for the course, excluding solo hours.

If you are on active duty, or training at less than 1/2 time, you receive the lesser of:

- The monthly rate based on the tuition and fees for your course(s); or
- The maximum monthly rate based on your training time.

Different benefits are available to veterans who served in the reserves during the post-Vietnam era. For information about these or other matters, write to the Veterans' Benefits Administration, Washington, DC 20420.

CHAPTER 3

WORKING CONDITIONS

PHYSICAL SURROUNDINGS

Working conditions for the drafter are generally quite good. As a drafter, you won't be spending your work week trudging through muddy fields or fighting the heat of a blast furnace, or enduring other uncomfortable situations. Instead, a typical drafting position will find you working in a clean, open space with good lighting and air conditioning. You will be seated in a comfortable chair before a computer or drawing board, with other equipment and supplies at your fingertips.

Of course physical surroundings vary, but these basics can usually be expected due to the precise nature of the work required. In the same way, while practices vary around the country and with different companies regarding such matters as wages, vacations, working hours, and benefits, some general standards apply with most drafting jobs.

HOURS AND BENEFITS

The length of the work week varies generally between thirty-seven and forty hours in a five-day week. Any work over this amount is overtime and is usually paid at one and one-half times the normal hourly rate. Work on Sundays and legal holidays is usually paid at double this rate. Certain holidays are allowed with pay, varying with local practices and traditions. Usually they are New Year's Day, Good Friday, Memorial Day, Independence Day, Labor Day, Thanksgiving Day, and Christmas Day. In some instances, as in certain union-management contracts, the employee is allowed a day off with pay on her or his birthday.

Paid vacations are the rule. The length of the vacation period usually depends upon length of service. A typical plan is the following.

An employee will receive paid vacations according to service acquired in each calendar year as follows:

> a) 1 week after 6 months' continuous service.
> b) 2 weeks after 1 year continuous service.
> c) 3 weeks after 10 years' continuous service.
> When a holiday occurs during vacation within regular weekly schedule, the employee is entitled to an additional day prior to or following her or his vacation period.

Health and welfare programs are common. Here, for example, is an excerpt from a union-management contract:

> A Group Insurance Plan, together with Group Hospital, Surgical, and Medical Expense Plans are made available on a contributory basis. The company shall be responsible for and bear the expense of the admini-

> stration of the plans and shall pay the balance of the net cost over and above the employees' contributions.

Provisions generally are made for sick leave. For example: "An employee who has completed six months of continuous service shall be granted up to 15 paid days of sick leave per year; nonaccumulative from year to year. In case of extended sickness, the employer may grant additional paid sick leave in accordance with past practices."

Another common feature of employment is some sort of retirement or pension plan. Generally, this is of the contributory type in which a stated percentage of pay is withheld from the drafter's pay check to be applied to a pension fund. If the drafter leaves, is discharged or is laid off, this accumulated amount plus interest is usually retuned to her or him.

Another employment practice that is becoming quite common is the allowance paid for layoffs, as when a plant or office shuts down, or work slackens and the drafter is unemployed through no fault of her or his own. Usually he or she will receive a layoff allowance that varies with length of service and may amount to one week's pay for six months of continuous employment and up to twelve weeks' pay for twelve years and over of continuous service.

A typical example of the job benefits being provided by most firms is represented in the following want ad.

> DESIGN—Draftspersons . . . The company job package includes good wages, such benefits as company paid life, medical insurance, paid holidays, vacations, savings and stock investment plan, retirement plan and educational assistance.

Another newspaper ad also gives a good indication of the availability of job benefits.

> Tool Designers . . . Excellent benefits—complete medical coverage (including Major Med.); dental plan; prescription drugs; life insurance, long term disability insurance, retirement fund; tuition aid, relocation expenses.

These ads are by no means exceptional. The employers, by and large, have found that it pays to have a satisfied drafter at work. It is an expensive business to have a large staff turnover; it means retraining, orientation, and other aspects of fitting a new employee into the particular systems at the plant.

Of course, all this does not by any means exhaust the list of job benefits provided by the industries of the country. Many firms provide contributions such as profit sharing. Other firms set aside a certain proportion of annual profits (if any) to be distributed among employees either as a yearly bonus or in weekly paychecks.

Another benefit that has become almost universal in industry is assistance in furthering the education of personnel. Employers may provide some or all of the following:

- Advance payment of tuition for job-related courses taken at colleges or technical schools.
- Reimbursement of tuition paid by the employee for job-related courses (may require successful completion prior to reimbursement).
- Payment for courses which are not directly related to one's job, but which are applicable toward a college degree.
- Book costs.
- Expenses for noncredit workshops and seminars, in some cases including travel costs.

Most firms large and small find it is good business to upgrade their employees' education, thereby providing the companies with personnel within their own ranks who can be promoted to better jobs.

The discussion of benefits on the preceding pages applies to the drafter in industry. Working conditions in civil service are covered by regulations such as apply to the federal civil service. They do not vary greatly from the conditions in industry. Paid sick leave, vacations, pensions, and so on may be somewhat more generous than those of private industry and there are no layoffs. On the other hand, the pay scales in industry are larger and the opportunities greater. However, there are some indications that the federal civil service may be moving to close the gap on pay schedules. In 1962, President Kennedy issued an Executive Order (#10988) which instituted formal and exclusive recognition of federal employee unions, the establishment of procedures for handling grievances and other problems, and the right of a qualified union to negotiate a mutually satisfactory written agreement with agency management.

Following this, Congress passed and the President signed Public Law 87–793, which embodies the principle of equal work, equal pay. Under this law, an annual survey is made by the Bureau of Labor Statistics to ascertain the wages paid in private industry for similar work performed in the government civil service agencies. This helps the government work toward developing pay scales that are comparable to those found in private industry.

REMUNERATION

In general, a drafter's earnings depend upon education, specialization, competence, and experience. Other factors are the type of company or employer for whom he or she works and the geographic location of the job. The drafter in private industry may be paid by the hour or on a weekly basis. There is no overall pattern, and conditions vary from job to job. As noted, he or she generally receives the usual fringe benefits, such as pension plan, paid vacations, sick and disability pay, and so on, but take-home pay reflects the deductions for some of these features, for federal and state income tax, and for social security. In considering pay schedules, therefore, it is important to remember that as much as 25 percent of a drafter's wages may be subtracted for these reasons, leaving the take-home pay that much less.

The United States Department of Labor conducts studies of average rates of pay in various occupations. In its publication, *Employment and Earnings Bulletin,* 1991, they reported that the median income for drafters in 1991 was $513 per week. This translates to a medium salary of $26,616 yearly. One-half of salaries studied were above this level, and one-half were below. For experienced drafters, earnings of $30,000 to $40,000 a year are not uncommon.

The bottom line is that persons employed in drafting occupations earn good salaries and benefits. Their income is higher than that of many other occupations, especially those where no specialized training or job knowledge is needed. In addition, the skills they use are applicable to other technical areas. Such skills can form the starting point for other careers for those who elect to pursue studies in engineering, architecture, inventing, or other areas where even higher salaries may be expected.

CHAPTER 4

GETTING STARTED

> "I didn't get the job because I don't have any experience. But how can I build up any experience if I don't have a job?"

You may have heard someone make such a statement, or wondered yourself about this kind of dilemma.

Fortunately, employers who hire drafters generally are not apt to follow such a philosophy. Because drafting skills can be built upon technical skills learned in school—whether vocational training in high school or a technical program at the college level—employers are often willing to hire an applicant who has had little or no previous work experience in the field. This is not to say that getting a job is necessarily smooth and easy, but with the right preparation your chances of landing a job can be quite good.

If you are graduating from high school after taking a course that included mathematics, science, and technical drawing, you have one great advantage in that you have at least the basic training required for a drafting career. An excellent start toward getting your first job is to consult with the guidance department in your high school. The guidance counselors usually keep in touch with neighboring centers of commerce and industry and often are called upon to recommend worthy students for job openings or for apprenticeship programs that may exist. The guidance people have your school record, probably know you personally, and should be able to help you, particularly if you tell them exactly what sort of a drafting job you are looking for. In this way, you can make an intelligent approach to appropriate employment. Don't miss your drafting teacher in this connection. Many of them keep in contact with industry and can also be of help.

Many vocational-technical high schools operate a cooperative course for students in senior classes. In this program, the school secures the cooperation of local industry or commerce, and the cooperating businesses employ the students according to their specialized school training on a work-study basis. A schedule is set up that allows the student to work a week on the job, where he or she gets actual work experience, and a week in school, where he or she continues with academic subjects. There are different variations of time schedules based on circumstances. If your school operates such a program, make sure you take advantage of it. This sort of cooperative education very often leads to a full-time job after graduation.

HELP-WANTED ADS

If such assistance is not readily available, there are still other methods of obtaining a job. One of the best ways is to consult the help-wanted ads in newspapers and trade journals. Check the ads in your local newspapers for the kind of beginning drafting job you could fill. If you live in a small town without many job openings listed, try to get newspapers from the nearest large city or metropolitan area. Your local public library will have a file of these daily publications as well as trade journals. Get in touch with all the promising leads. This usually must be done by a letter of application.

Watching these ads often brings results; here are some ads clipped from Newark, New Jersey; New York City; and Boston, Massachusetts, newspapers:

> *Drafters*—Permanent Positions in Home Office—Electrical or Mechanical Beginners—with Drafting school Training or equivalent.
>
> *Drafting Trainee*—Knowledge Mechanical Drawing—learn layouts for Die Making.
>
> *Drafter, M/F.* Position available for bright person willing to learn design. Must have drafting education—will train the right person.
>
> *Junior Drafter—M/F*—Electronics company seeks person for electrical drafting—must have high school drafting and have some knowledge of electronics—Full time, permanent position.
>
> *Drafter*—opportunity to learn progressive die layout and programming of N.C. controlled machines (automated).

You probably have had some training in high school in the way to write application letters. Don't forget these lessons. Write or type a legible, logically arranged letter on clean 8 1/2-by-11 inch bond paper. In your letter give the usual information as to schooling, trade training, abilities, and so on. Don't claim anything you can't prove; if you have had no actual industrial drafting experience, don't claim any. However, don't be shy about your ability and potential. Try to find out about the product of the firm so that you may be able to tell the potential employer how you think you can fit in. Do not enclose original documents such as diplomas, certificates of merit, or school transcripts—send copies of any that are required. In short, while you should not make unsound claims, you should call attention to your capabilities, training, and ability to fit into a beginning job and to grow with the opportunities offered.

EMPLOYMENT AGENCIES

Next to the want ads in newspapers, the best sources of employment opportunities are public and private employment agencies. Public employment agencies are maintained in all states. In fact there are about 1,800 such offices throughout the nation. If you don't know where the nearest one is located, look it up in the phone book or write for information to the employment service in your state capital. Many of these employment agencies also offer counseling services.

In addition, there are reputable private employment agencies located in most cities. These have access to many job opportunities, and it certainly is worth your while to take advantage of their help. However, some private agencies require you to pay a fee, which may be as much as a week's salary, if they find you a job.

Other Job Lead Sources

Other possible sources of job leads are local unions or labor organizations. Many of them have a fund of information about local job opportunities that can be of help. Perhaps one of your family members or friends can put you in touch with these offices. And while you are at it, don't be shy about telling your relatives and friends about the kind of job you are seeking. Often a good lead can be developed this way.

CIVIL SERVICE

If you are interested in a civil service career, consult your municipal, county, or state service commission for details. For the federal civil service, write to the Office of Personnel Management, Career Entry Group, 1900 E. Street, N.W., Washington, DC 20415.

THE INTERVIEW

Suppose that your efforts have paid off, and you have been asked to come in for an interview by a prospective employer. Before you go, try to learn something about the company so that you can have an intelligent approach to its requirements and how you can best fit in with them. Perhaps you know someone who is employed at the particular place. A talk with this person may provide you with valuable information.

As for the interview itself, some pointers may prove of value. Undoubtedly you already have information as to the proper conduct of a job interview, probably provided by guidance counselors and others. However, repetition can do no harm, and it can mean the difference between getting a good starting job and just floundering around in a maze of difficulties. Incidentally, studies of the causes for employee dismissals have put inability to adjust to people or to get along with others as the most important reason; inability to perform or to do the job as such has been identified as a secondary cause.

The New York State Department of Labor published a free pamphlet entitled *Why Young People Fail to Get and Hold Jobs*. It listed as important factors the following:

Your Appearance—can be the difference between getting the job and getting the "brush-off."

Attitude and Behavior—play almost as important a part in getting and holding a job as does skill.

Ignorance of Labor Market Facts—can result in costly mistakes.

Misrepresentation—is bound to be discovered and work to your disadvantage.

Sensitivity about a Physical Defect—can be a serious obstacle to getting and holding a job, if you let it.

Unrealistic Wage Demands—and the odds are you won't get the job.

Absence or Lateness Without Good Reason—can cost you your job and could make it difficult to get another one.

Insufficient Training—is an obstacle to getting the job you want.

Insistence On Doing the Job Your Way—most likely will create animosity and work to your disadvantage.

Balk at Entry Requirements—and you are likely to miss your big opportunity.

Apply for a Job with a Friend Along—and you probably won't be hired.

Be sure to dress properly when you go for the interview. This is not idle counsel, for first impressions are very important. Many an applicant has failed to land a job because of an unkempt or slovenly appearance. Don't chew gum. Bring your social security card with you, as well as samples of your work and other documents that may be required. When you meet the interviewer and he or she offers to shake hands, use a moderately firm grip. Either a crushing or a limp handshake may seem to indicate a lack of control or enthusiasm. In speaking with the interviewer, try not to be scared. On the other hand don't be brash and don't make extravagant claims you won't be able to prove.

Many companies require job applicants to take various aptitude and psychological tests. You probably have taken such tests in high school and should therefore know what to expect. At any rate, if you do have to take tests, don't let it upset you; just turn in your best performance. Some of these tests can be of help in indicating your own personal potentials.and capabilities.

The following is a list of pointers on how to apply for your first job. This list is from the guidance department of a large vocational-technical high school in New York City.

Pointers on How to Apply for that All Important "First Job"

Whether you are applying in person, over the phone, or by mail, you will want to put your best foot forward when applying for that *first job*. You will want to make the best possible impression on an employer. This means being prepared to present yourself and your qualifications effectively.

Fortunately, or unfortunately, first impressions during the interview *do* count. Prospective employers will judge by what they see. It is important that you:

- Be neat and well groomed
- Check to be sure that your clothes are clean and pressed
- Have your shoes polished and in good condition
- Have your hair and fingernails neat and clean
- Watch your grammar and usage at all times, avoiding current fads of pronunciation and slang
- Let the employer see that you are truly interested in the job for which you are asking to be considered
- Show some enthusiasm for the job and the company
- Take aptitude tests and complete application forms willingly and cheerfully
- Do not be too modest! Inform the employer of your special abilities, skills, and interests
- Do not be afraid to ask questions and respond politely and honestly to those asked of you
- Do not be afraid to smile occasionally
- In a "nut-shell" try to relax and be yourself
- When the interview is over, thank the interviewer and walk out.

In many instances it is wise to follow up the first interview with a thank-you letter, and after the lapse of several days, with another personal call. You must convince the employer that you are really interested in working for the company. Be *brief at this time* because your only purpose is to keep yourself in the employer's memory.

If you apply by letter or telephone, remember that you need to impress the employer with your qualifications so that he or she will want to consider you further. Your telephone voice and manner or your neatly typed, well-worded letter will give a clear picture of you. (Correct spelling is important.)

The approaches suggested for finding a beginning job as a drafter can be applied even if you are not a graduate of a high school with a drafting program. If you have had adequate training through a good home-study course or by attending an evening program at a local school or other part-time school, you can try for a job by the various methods outlined and especially by consulting and using the services of the training institutions whose courses you have taken. Of special value in this case are the public and private employment agencies, who can guide you to apply to employers who are most likely to be flexible in considering different kinds of training.

You will also find that a course in computer-assisted drafting will be of special value. If your training took place before this equipment was available, you will also find this course valuable in updating your skills.

CHAPTER 5

ANALYSIS OF JOB QUALIFICATIONS

As with any other specialized field, drafting is not for everybody. To succeed in drafting—and to be happy with the work—you not only will need the right education, but also should possess certain aptitudes. Of course there is no magic formula for success, and your own desire to do well is the most important factor. In general, however, most persons employed in drafting will demonstrate certain basic qualities.

First of all, you should like drawing and be reasonably good at it. Even with the advent of the computer, the ability to sketch out details will be important. Even more importantly, you should be detail-oriented and observant, for attention to details is at the heart of drafting work. Other desirable qualities include being comfortable with mathematics, unafraid of computers, patient, and neat. You will also need the capability of sitting behind a desk for a long period of time without becoming stir-crazy. If you feel that you have these qualities, drafting may be well suited to your personality.

It is important to remember that the term *drafter* covers a great variety of specialized occupations. While these require the same basic drafting techniques, each one is sufficiently different to need years of specialization for fullest proficiency. Merely listing some of them indicates these diversities:

Architectural Drafter
Aeronautical Drafter
Autobody Drafter
Marine Drafter
Mechanical Drafter
Civil Drafter
Patent Drafter
Tool Design Drafter
Heating and Ventilating Drafter
Hull Drafter
Map Drafter
Topographical Drafter
Structural Drafter
Geophysical Drafter
Sheet Metal Drafter
Technical Illustrator

An analysis of some of the main divisions of drafting jobs follows as an aid in determining preferences.

MECHANICAL DRAFTER

Probably the largest group of drafters in the United States is the *mechanical drafters*. These technicians work on the design and detail of special machinery and machine tools of all descriptions: production tools, such as dies, jigs, fixtures, and gauges; automation equipment, such as loaders, chutes, bucket conveyors, and so on; transportation equipment and vehicles, such as automobiles, locomotives, freight cars, and so on; plant equipment and layout; power plants and equipment; military hardware, and the like.

Included under the general heading are such titles as *tool design drafter,* or one who works on production tooling and may be a specialist on die design, fixture design, jig design, or any other branch of mass production tooling.

Of all the drafting specialists, perhaps that most closely connected with the production process or the shop is the *tool designer* or *drafter.* Mass production of metal goods of every description is dependent upon such devices and tools as dies, jigs, fixtures, gauges, cutting tools, and machine attachments. With the help of one or a combination of these devices, it is possible to produce a particular part in great numbers at a high speed, each of them exactly the same and meeting closely held specifications and tolerances. Parts thus produced can quickly be assembled to form automobiles, typewriters, flashlights, telephones—any of the innumerable objects with which we are so familiar. The tool designer must not only be an expert drafter but must also have a great deal of experience in and an extensive knowledge of machine tools and machine shop methods. He or she usually works with the tool engineer in selecting the machines and processes for doing the particular production job and then proceeds to design the necessary tools, which often calls for a high degree of inventive ingenuity. The tool drafter is thus a combination of expert mechanic, skilled drafter, and ingenious innovator—and thus a key figure in the production process.

Another large subdivision of drafting specialists is the *product or engineering drafters,* who are concerned with the design and detail of the many products that pour out of our factories. These products include measuring and weighing devices, air conditioners, record cutters, and business machines. Of course, in time a drafter engaged in product design may become a specialist in the particular field. Good examples of specialization can be found in the automotive industry. A drafter may specialize in the design and detail of the engine or the design and detail of the various accessories, each of a highly technical nature. A very interesting job is the one done by the *auto body drafter,* who is part of a team that sets the style and design of the automobile body. Some product designers are in highly complex modern technologies, such as the design and detail of atomic weapons, missiles, rockets, and space vehicles.

As an example of how the product drafter and the tool designer work, let us suppose that a company that manufactures typewriters has, after much careful calculation and design, produced a new and advanced model. Several copies or prototypes have been made by hand and have been thoroughly tested from every aspect. Changes have been made as indicated by this severe testing, and the model now is ready for production.

During this time, the product drafter, in consultation with the engineers and designers of the new model, has been producing layouts, detailed drawings of every part, subassemblies, assemblies, and other drawings as required. As the machine is perfected, each detailed part drawing is changed to suit the new specifications until finally a complete, corrected, and checked series of drawings is ready. These show shapes, sizes, materials, hole locations, and all the other information necessary for production.

At this point the tool engineers will examine each detail or part drawing and will decide how it is to be manufactured; in what sequence; what machines are to be used; and what fixtures, dies, jigs, and tools are necessary. He or she will also supply any other information pertinent to production. Now the tool drafter is called in, and from these production schedules he or she will design or lay out the tools indicated for each part. It may be a jig to hold the part for drilling or tapping, a fixture to hold it in the milling machine while cuts are being taken, a part to hold it in an automatic sequence for various machining processes, or a die to stamp and form other parts. At any rate, the tool drafter must be able to make the drawings that will be used for making the tools, devices, and fixtures for the mass production of each part of the typewriter as drawn by the product drafter.

Patent drafters are another specialization. They are masters of their trade who have become specialized in the area of patents. Working with patent attorneys or directly with inventors, they transfer ideas for new inventions into patent application drawings in accordance with rigid standards prescribed by the United States Patent Office.

The machine drafter or *designer* works on the design and detail of special automatic machinery and machine tools. This is one of the most challenging branches of the trade since it calls for a profound grasp not only of mechanical technology but also of mechanics, electronics, air and hydraulic controls, and metallurgy. A high degree of ingenuity, imagination, and forethought also is needed. Perhaps an example will make this plain.

A quantity baker has perfected a new brand of bread, and it is to be put on the market. Part of the new equipment required is a high-speed packaging machine to use a newly invented plastic wrapping material. This problem is placed by the engineers before the machine designer, who proceeds from the depths of experience to sketch up various methods of coping with the job. After one particular method is selected, the machine designer will then proceed to lay out or draw to scale the assembly of such a packaging machine. Detailers will then pick off and make working drawings of the subassemblies and the various parts or members for the shop where they will be made.

Such mechanical marvels as cigarette- and cigar-making machines, automatic machine tools, and automatic bowling alley pin spotters were all produced in a similar manner by machine designers and detailers working with engineers and other technical personnel.

ARCHITECTURAL DRAFTER

The *architectural drafter* develops the ideas of the architect into working drawings for construction. When a residence, an office building, a factory, a hospital, or any other edifice is being planned, the architect will submit to the client a number of ideas in the form of perspective sketches or pictures showing the completed building as well as sketches of proposed floor plans and elevations. When a design has finally been selected, the architect will turn it over to his team of architectural drafters to develop the floor plans, sections, and front and side views, or elevations, of the proposed building. In addition to the structural features, the drafter will also draw in the numerous items that will eventually go into the completed building. These include the heating, ventilating, plumbing, and electrical systems. He or she will also provide for such services as elevators or escalators and such additional details as specifying the materials to be used. When all these details are completed and approved, the drawings can then be turned over to the contractors for bidding and will be used for erecting the structure. Under these conditions, the architectural drafter becomes an expert in building construction and may also develop the ability to make perspective drawings or renderings of the completed building as well as its details.

Allied to architectural drafters are specialists in structural iron work known as *structural drafters*. These people lay out the plans for bridge trusses and other bridge components, for steel building frames, for steel towers used in high tension electrical transmission lines or for other purposes, and any other structures that depend on steel-and-concrete walls, trusses, girders, columns, and so on. These drafters become expert in calculating the stresses and strains on structural components and structural design.

Other building construction specialists are the *heating and ventilating drafters,* who lay out installations for these purposes; *plumbing drafters,* who plan the installation of sanitary and water services; and *electrical drafters,* who prepare the working plans and wiring diagrams for electrical equipment.

LANDSCAPE DRAFTER

The landscape drafter works with the landscape architect. The latter is concerned with the planning and designing of landscapes and with projects such as recreational areas, parks, airports, highways, parkways, private estates, and the like. The landscape drafter, to paraphrase the *Dictionary of Occupational Titles,* works with the architect by preparing detailed scale drawings and tracings from rough sketches and other data provided by the architect, performing duties common to the drafter who may prepare separate detailed site plans, grading and drainage plans, lighting plans, paving plans, irrigation plans, and drawings and details of garden structures. The drafter may build models of proposed landscape construction and prepare colored drawings for presentation to clients.

MARINE DRAFTER

This specialized branch of drafting has to do with the mechanical and structural design and detail of ships of all kinds from small boats to passenger liners and from submarines to warships, as well as other marine equipment and structures, such as floating docks, piers, and so on. The marine drafter concerned with the mechanical equipment of ships, such as boilers and engines of all descriptions, steering gears, auxiliary ship equipment, and piping, is called a *ship engineering drafter.* A *hull drafter* makes the designs and details of the hull construction of a ship and its related parts, which is an art in itself and involves specialized drafting techniques for laying out complex curves. There are other specialists covered by the general term of *ship detail drafter,* who work on the electrical circuits, plumbing systems, and other details of ship design.

TOPOGRAPHICAL DRAFTER

Topographical maps are made by the *topographical drafter* from the notes and sketches of land surveyors, aerial photographs, and other available sources. He or she is a specialist in fine lettering and in using symbols to depict geographical features, such as mountains, rivers, and depressions of the area being mapped.

TECHNICAL ILLUSTRATOR

Another type of specialized drafter is the *technical illustrator.* He or she is an expert in mechanical drawing and in all methods of pictorial representation, including isometric drawing and perspective drawing. This drafter uses many of the techniques of the artist, such as the air brush for shading and bringing up highlights on drawings. He or she takes the detail drawings of a device and assembles them into a picture of the complete article, which is then reproduced for use in instruction manuals or advertising material. All the parts of the assembled device must be pictured in relation to one another in section or in an exploded assembly for the guidance of persons making or using this device. For example, every auto mechanic is familiar with the instruction manuals issued by the automobile manufacturers with every model put on the market. These show all the parts of a car in detail, in subassembly, and in relation to one another, and are an invaluable aid in maintenance and repair work.

There are also many other specialized areas of the drafting trades, too numerous to include here. Enough has been said, however, to indicate that there are plenty of specializations to satisfy every talent or interest. It is true that very often the first job of a beginning drafter will dictate the field of specialization, but if an individual desires a change, there is a place for her or his particular talents somewhere in the field of drafting.

CHAPTER 6

ADVANCEMENT—RELATED FIELDS

Whenever you think seriously about any career field, a key consideration should be the possibility for advancement. After all, your working life will most likely extend for forty years or more, and you probably will not want to spend all that time at the same job level. Instead, you will want a chance to grow and develop in terms of authority, responsibility, and rewards.

Fortunately, positions in drafting tend to offer good possibilities for advancement. Persons starting out as novice drafters often work their way up to supervisory positions, including the highly desirable post of *head drafter.* It is not uncommon, in addition, for drafters to combine job experience and additional training to advance to other fields such as engineering, architecture, or teaching, or to form their own business and get ahead that way.

Such famous persons as George Westinghouse, who founded the Westinghouse Corporation; Charles Kettering, one of the pioneers of the automobile industry; and noted architect Edward Durrell Stone all started out as drafters. Many other persons have gone on to all kinds of rewarding careers after starting out in drafting. Following are some of the possible career avenues that may be followed by the drafter who has achieved a solid record of experience and capability.

HEAD DRAFTER

The top job in the drafting field is that of *chief* or *head drafter.* This is largely an administrative and supervisory job in which, as head of the drafting department, the chief deals with engineers, architects, scientists, technicians, or whatever professional personnel the particular area requires. The head drafter lays out the drafting work at hand to suit time schedules, assigns the work to the design and drafting personnel of the office, and supervises and otherwise follows all this through checking to completion. He or she may hire and discharge and may have to deal with unions and the like. The head drafter probably sits in on high-level conferences, as he or she has a detailed grasp of particular technical areas and possible design solutions as well as of administrative detail. As can be seen, this is usually a highly responsible post, but one into which a competent drafter and designer can develop.

ENGINEER

For most of the drafting specialties—*mechanical, structural, marine, civil, electrical,* and *electronic*—another logical road for the development of the drafter is that which leads to the profession of engineer. This road is a long and arduous one. It requires many years of study leading to a technical degree in engineering courses from an accredited college or university. That it can be done is attested to by the thousands of young men and women who have attained professional standing by this method. In many cases, the companies that employ drafters encourage such ambitions by paying course tuition.

Typical of this is the approach used by the General Electric Company for ensuring advanced training for those of its technicians who wish to work toward technical degrees. This is known as the Tuition Refund Plan. Under its provisions and with prior managerial approval, any employee may receive a refund of part or all of the tuition expense incurred in taking undergraduate or even graduate education. The employee must maintain good standards and receive satisfactory grades in these courses. This plan has proven to be advantageous both to the employee and to the company. Many firms realize the advantages to be gained by paying for and encouraging the further education of their own employees. For many ambitious drafters, this type of plan has provided the means for securing advanced training and, eventually, professional degrees.

Prospects for engineers have never been better. There exists a constant and growing demand for these professionals in all branches of industry and government. A glance at the help-wanted sections of metropolitan newspapers, at professional publications, and at the literature from government agencies would confirm this. With the expansion of such new technologies as atomic energy and electronics, demands for engineers will continue to grow.

LICENSURE

All the states license professional engineers. In fact, in many states this is a prerequisite before an engineer is allowed to operate independently. For example, building plans in many states must carry the signature and seal of a professional structural engineer for the steel work and a professional mechanical engineer's signature and seal for mechanical equipment. The drafter who has attained an engineering degree can enhance his or her livelihood by taking the necessary examination for the professional engineer's license. In New York State, for example, a candidate must show at least four years of satisfactory experience in his or her specialization, in addition to graduation from a registered college or school of engineering. However, if one can show twelve years of practical experience in engineering work of a satisfactory grade and character, this college degree is not required.

ARCHITECT

In a similar manner, an architectural drafter can head for the goal of becoming a registered architect. In general, the requirements include graduation from a recognized professional school of architecture followed by three years of practical experience in an architect's office. However, in most states it is possible to substitute from ten to twelve years of practical experience for the architectural school training and subsequent experience. An architectural drafter can thus obtain a registered architect's license by passing the required tests after the necessary period of experience. He or she can then hang out a shingle as a practicing architect. Through hard work and study, therefore, it is possible for a drafter to arrive at a professional standing and perhaps open an independent office.

However, for the person who is not heading for a professional career, there are other ways to advance. A drafter is in contact with so much of the technical development as well as the product design, production tooling, and machine design in a particular firm that opportunities of growing into related and possibly more rewarding positions often present themselves.

RELATED FIELDS

Because of an intimate knowledge of the products of the company that employs her or him, the drafter can often act advantageously as the company's sales and service representative. Or the drafter can work as its *field erector* or assembler, going into any locality to direct the erection and servicing of large machinery and equipment. Branching out into various other manufacturing positions, such as *production control,* is possible, as is specializing in *methods, tool engineering, quality control, time and motion study,* especially if he or she has taken the pertinent technical work in extension courses. The list of such opportunities is almost endless.

Similarly, an architectural drafter can in time become a *specification writer,* a highly technical and well-paid occupation in which he or she prepares in the form of a pamphlet or book the exact specifications of all the materials that go into a particular building, structure, or repair job.

Or the drafter may develop into a *construction supervisor,* in which position he or she will supervise the job in the field to ensure that the building goes up on time and in accordance with plans, specifications, and the local building code.

For the drafter who does not become a professional engineer or architect, yet wishes to get away from the board or computer and branch out into an allied and perhaps more profitable field in a particular technology, the list of opportunities is endless. These opportunities depend upon job conditions, the drafter's experience and personality, educational and technical background, and ability to see and be in a position to take advantage of the chances that come along or that can be created.

YOUR OWN BUSINESS

It is possible for a drafter to go into her or his own business. Organizations exist that offer design and drafting services for almost all branches of industry. By combining forces, several designers can do this on a profitable basis, offering these services to companies that need them but have no drafting rooms of their own, or that need special work and do not wish to expand their existing drafting facilities. This type of venture requires considerable trade experience, many good contacts, and business acumen on the part of those establishing such a business. But it does not require as large a monetary investment as would be needed in setting up a machine shop or a manufacturing company requiring large amounts of machinery. All that is needed is a large loft or office, drawing boards and equipment (including a computer and drafting software), and drawing supplies. Blueprinting and duplicating services can be supplied by firms specializing in these areas. The federal government uses a great deal of such drafting services, and there are subcontracts to be had from larger contractors doing business with the government.

Sometimes in the course of a career, a drafter may design or invent a new item that can be marketed to advantage. However, it should be kept in mind that anything a drafter invents or designs during working time usually belongs to the employer. At any rate, there are new areas in many technologies that offer such opportunities—for example, in electronics, where new products appear constantly. In fact, many large and successful concerns specializing in electronics are only a few years old, and many began with little more than an advanced idea. The same can be said for many new industries centering around the uses of atomic energy and other new scientific discoveries. In short, a drafter with some inventive or creative ability can find the opportunity to establish a business, making and selling some new product fostered by our technological advances. Of course, this involves many considerations of financing, investment, manufacturing, accounting, and marketing—considerations that must be taken into account if a successful business is to be established.

Another factor that one must be prepared to confront is the matter of getting a patent or patents on the new product to be manufactured. This is by no means a simple matter. Obtaining a patent can be a costly, lengthy, frustrating experience with many pitfalls. It may take up to three years before the coveted patent is finally obtained, and then it is up to the inventor to make it pay and to guard it against infringements, which are sure to come if the product is a commercial success. However, if it seems worthwhile to pursue the project, don't try to do it all yourself. Take it to a reputable patent attorney with the necessary experience, connections, and resources to make the required search through patent records to make sure the idea has not already been patented. The U.S. Patent Office in Washington, DC, keeps a list of registered attorneys to which you can refer. Make sure your attorney will help you write up the invention in the technical language required by the Patent Office, will have the necessary drawings made by a patent drafter in accord with official standards, and will finally submit the whole thing to the Patent Office and follow it through to, it is hoped, a successful conclusion. It is also a good idea to retain the attorney for the legal proceedings that may follow as the pirates move in to infringe on your patent if your invention gives promise of becoming a commercial success.

For more detailed information write to the Commissioner of Patents, Patent Office, Washington, DC 20025. The following books can be of help:

- *Invention and Design.* McGraw-Hill, 1985.
- *Inventing for Fun and Profit.* San Francisco Press, 1990.
- *Marketing Your Invention.* American Bar Association, 1989.
- *So You've Got a Great Idea.* Addison-Wesley, 1986.
- *The Elements of Invention.* Macmillan, 1990.
- *The Inventor's Handbook: How to Develop, Protect and Market Your Invention.* Betterway Publications, 1990.

An aspiring drafter can turn to many sources for help. One of the most helpful of these is the Small Business Administration, Washington, DC 20416, which has field offices throughout the country. It was established in 1953 by the passage by Congress of the Small Business Act. It provides help to small business enterprises in three broad areas including financial assistance, management and technical assistance, and help in obtaining government contracts. The Small Business Administration will loan a small firm up to $150,000 on a regular business basis, and on certain conditions it will cooperate with banks in extending loans to such companies. Another of its devices for extending financial help is the National Association of Small Business Investment Companies, 323 W. 8th St., Kansas City, MO 64105, which will allow capital equity, debt financing, and consulting services under certain conditions.

The Small Business Administration has close contacts with several government agencies such as the Defense Department, the National Aeronautics and Space Administration, and others. All of these place large contracts, but under federal regulations such agencies must give small enterprises first consideration in bidding for subcontracts.

Factors of Failure and Success

Before any individual decides to go into business, however, it would be wise of that individual to consider the factors that make for success in such a venture and, conversely, those factors that lead to failure. One analysis cites the major reasons for failure as follows: neglect, 2.3 percent; fraud, 0.9 percent; economic conditions, 5.0 percent; lack of experience in the line, 10.2 percent; lack of managerial experience, 13.5 percent; unbalanced experience,

18.6 percent; incompetence, 47.8 percent; disaster, 2.7 percent.

Of the causes listed, poor management and personal attributes are by far the largest factors in the failure of small businesses. It would seem, therefore, that the drafter aspiring to his or her own business would do well to prepare by getting adequate training through appropriate courses at a college, university, or other educational center, or through extension courses.

The following list includes several references that can be of help:

- *How to Organize and Operate a Small Business.* Prentice-Hall, 1988.
- *How to Start Making Money in a Business of Your Own.* Gordon Press, 1987.
- *The Complete Entrepreneur.* Mercury Books, 1991.
- *The Small Business Legal Advisor.* McGraw-Hill, 1991.
- *Small Business: An Information Sourcebook.* Oryx Press, 1987.

TEACHING

Becoming a teacher can be a rewarding career. There are many advantages in teaching: job security, paid vacations in the summer as well as holidays, pensions, sick-pay benefits, and a shorter working day than in industry. It may be true that beginning salaries are somewhat lower than the journeyman drafter can get in the industrial world.

Qualified drafting instructors may find positions in secondary schools, vocational-technical high schools, technical institutes, and community colleges and universities. It is often said that no one can become wealthy as a teacher, but it can be a worthwhile career with many gratifying and rewarding facets. If you are interested in teaching, get in touch with the board of education in your city and with local institutes of higher education, technical schools, and the like.

CONCLUSION

A person beginning as a drafter can find many opportunities for broadening and enhancing life and livelihood. Because of the strategic nature of the occupation, which is the vital link in the chain of production between the idea and its manufacture, the drafter is often in a position to deal intimately with new ideas, new materials, and new technologies that offer many opportunities. There are endless paths to follow from the drafting board or computer to greater opportunity. The opportunities mentioned earlier by no means represent all the possibilities. They are, however, good indications of what can be achieved.

CHAPTER 7

LABOR AND OTHER ORGANIZATIONS

One of the major factors influencing American history over the past one hundred years has been the development of labor unions. Prior to that time, laborers had little control over the working conditions, all real control being exerted by those who owned the business. Since the advent of unions, however, a whole new era of workplace democracy has developed.

Although unions as we know them did not evolve until the nineteenth century, as early as 1648 working people were beginning to form craft organizations such as those formed by shoemakers and by barrelmakers (coopers) in Boston. These as well as other craft groups for some time came together mainly for benevolent, financial, social, and burial assistance.

While such groups were not unions as we know them, they did provide the training and experience needed for the later formation of craft guilds that first experimented with many of the approaches now associated with unions. For example, in 1799 the Philadelphia shoemakers were the first to hold meetings where employers and workers met and discussed such issues as higher wages, shorter hours, the closed union shop, and other matters of prime importance at the time. While no great improvements were effected, this did bring to public attention the need for changes that in time were made, and it encouraged similar meetings in other major cities of the day.

Other techniques, now familiar, were developed during the period between the Revolutionary War and the Civil War. The first authorized strike was called by the printers of Philadelphia in 1786 and was soon followed by strikes by shoemakers in the same city and in New York City. But as the unions, learning by experience, improved their methods for enforcing their demands, particularly for wage increases and shorter hours, the employers reacted by using strike-breakers and by bringing the striking unions into court. There they were made answerable to an old doctrine of English law that declared that combinations of workers to raise wages could be illegal as a conspiracy against the public. The conspiracy doctrine was to hamper the unions of this country into the 1900s.

The fortunes of unions waxed and waned through the nineteenth century in accord with the economy. But no really effective organizations were achieved until 1881, when under the leadership of Samuel Gompers and Adolph Strasser, both cigarmakers, six large craft unions—the printers, iron and steel workers, molders, cigarmakers, carpenters, and glass workers—with a number of lesser groups formed the Federation of Organized Trades and Labor Unions. In 1886, the American Federation of Labor (AFL) was formed on the wreckage of the old Knights of Labor, which had been predominant until then. When the new Federation of Organized Trades and Labor Unions merged with the new AFL, the AFL that we are now familiar with was born under the leadership of Samuel Gompers. This powerful group was able to obtain major improvements in working conditions and to get favorable labor legislation through Congress during the period between 1886 and the beginning of World War II.

After World War II, the labor unions of this nation experienced a considerable growth in membership, influence, and strength, culminating in the merger of the American Federation of Labor and the Congress of Industrial Organizations into one great union. Despite all this, the so-called white-collar workers, including office workers, drafters, technicians, and the like, have proven difficult to organize. Drafters, who are proud of their professional status, have traditionally been disinclined to join labor federations. The many reasons include the presence of a large number of individualists among the practitioners of the trade and a constant fear of "regimentation." However, some drafters have been successfully organized either into affiliates of national labor federations or into local plant unions.

It may be that a plant where you obtain work as a drafter is unionized with either a CIO type of organization in which all the plant workers are members, regardless of job classification, or with the AFL type of craft unions. In this case, you may be required to join the labor organization, and dues will probably be taken from your paycheck. The union then is your agent in labor negotiations with management, and you should do your part in making sure it is a good representative. Attend the meetings, take part in the debates and proceedings, and in general participate in its affairs. If you don't do your share, you have no right to grumble if affairs go badly or if the wrong kind of person dominates the union and its activities.

There are several national labor organizations concerned with drafters. The largest is the American Federation of Professional and Technical Engineers, AFL-CIO, with headquarters in Silver Spring, Maryland. This union has been in existence since 1918 and has locals in all parts of the country. It also has connections with the Canadian Labor Congress. To quote from its Constitution:

> **Article I**
> Section 1. This Federation shall be devoted and dedicated to organizing into a national organization and shall embrace within its jurisdiction, all individuals who work in the general technical and scientific field of engineering, architecture, and allied occupations for the purpose of representing them in collective bargaining and otherwise improving their economic status and conditions of employment.
>
> **Article IV**
> Section 1. Any individual employed or qualified for employment in the general field of engineering or architecture shall be eligible to membership in this Federation upon application, election, and initiation in the manner and form prescribed in this Constitution. Eligible membership in the Federation shall include, but is not limited to, persons in such occupations as aeronautical, chemical, civil, commercial, electrical, mechanical, mining and metallurgical, sales, structural and research engineering; architects, draftsmen, metallurgists, physicists, engineering inspectors, test technicians, planners, estimators, specification writers, technical clerks, time study men,

blueprinters, engineering and laboratory assistants and aides, etc.

Under the provisions of the Taft-Hartley Act and the Landrum-Griffin Act of 1959, it is possible for the majority of workers in a drafting office to select an organization such as the one above to act as their bargaining agent with the employer. Or they may form an independent local unit of their own, with no national affiliation, to act in that capacity. The laws also carry provisions for the settling of labor grievances and the general procedures to be followed in labor management negotiations.

PROFESSIONAL ORGANIZATIONS

In addition to these labor agencies, there are professional organizations that can be of help to the drafter. The largest of these are the engineering societies which share a headquarters building in New York City at 345 East 47th Street. Among them are the American Society of Mechanical Engineers, the American Institute of Electrical Engineers, the American Institute of Chemical Engineers, and the American Society of Civil Engineers. All have their central offices at this address. They maintain a fine technical library, an information service, and an employment agency for professional and technical people. While interested primarily in professional personnel, the societies make provision for the memberships of designers and top drafting people. Architects, too, are organized into such societies as the American Institute of Architects with offices in Washington, DC.

Management, as well as labor, has its organizations. The National Association of Manufacturers, which is comprised of a broad cross section of the country's manufacturing companies, is probably the most representative of such organizations. In addition, practically every branch of industry has its own trade organization, such as the following: American Institute of Steel Construction, Brick Institute of America, Canning Machine and Supplies Association, American Management Association, National Tooling and Machining Association, and Machinery Dealers National Association. All of these groups can be of help supplying information, figures, and sometimes employment counsel in their own fields.

As an aid in securing information and possibly helping toward employment or better opportunities, the address of the various organizations are listed as follows:

American Association of Community Colleges
1 DuPont Circle
Washington, DC 20036

International Federation of Parks and Technical Engineers
818 Roeder Road
Silver Spring, MD 20910

American Institute of Architects
1735 New York Avenue
Washington, DC 20006

American Institute of Chemical Engineers
345 East 47th Street
New York, NY 10017

American Design Drafting Association
P.O. Box 799
Rockville, MD 20848-0799

Institute of Electrical and Electronics Engineers
345 East 47th Street
New York, NY 10017

American Society of Civil Engineers
345 East 47th Street
New York, NY 10017

American Vocational Association
1410 King Street
Alexandria, VA 22314

U.S. Council for Energy Awareness
1776 I Street
Washington, DC 20006

Chamber of Commerce of the United States
1615 H Street, NW
Washington, DC 20062

Accreditation Board for Engineering and Technology
345 East 47th Street
New York, NY 10017

Federal Civil Service Commission
Washington, DC 20415

National Association of Manufacturers
1331 Pennsylvania Ave., NW
Washington, DC 20004

Career College Association
750 First Street NE, Suite 900
Washington, DC 20002

National Home Study Council
1601 18th Street, NW
Washington, DC 20007

Association for Manufacturing Technology
7901 Westpark Drive
McLean, VA 22102

National Small Business United
1155 15th St., NW
Washington, DC 20036

National Tool, Die and Precision Machinery Association
9300 Livingston Road
Ft. Washington, MD 20744

National Society of Professional Engineers
1420 King Street
Alexandria, VA 22314

Shipbuilders Council of America
1110 Vermont Ave., NW
Washington, DC 20005

Society of Manufacturing Engineers
P.O. Box 930
One SME Drive
Dearborn, MI 48121

Society of Women Engineers
345 East 47th Street
New York, NY 10017

U.S. Department of Education
Division of Vocational Education
Washington, DC 20202

APPENDIX A

SUGGESTED READING

The following periodicals offer current, challenging articles that will give you ongoing information about the field. Most are available from public or technical libraries, or, you can write to the addresses provided.

CAD/CAM Abstracts
R.R. Bowker
249 W. 17th Street
New York, NY 10011

Chilton's Automotive Industries
Chilton Publishing Co.
Radnor, PA 19089

Computer Design
PennWell Publishing Co.
1 Technology Park Drive
Westford, MA 01886

Computer Graphics
Association for Computing Machinery
11 W. 42nd Street
New York, NY 10036

Computer Graphics Review
Intertec Publishing
730 Boston Post Road
Sudbury, MA 01776

Computer Graphics World
PennWell Publishing Co.
1 Technology Park Drive
Westford, MA 01886

APPENDIX B

UNITED STATES OFFICE OF PERSONNEL MANAGEMENT REGIONAL OFFICES

Regional Headquarters	*Area Served*
Federal Building, Suite 904 75 Spring Street, SW Atlanta, GA 30303-3109	Alabama, Florida, Georgia, Mississippi, North Carolina, South Carolina, Tennessee, and Virginia.
William G. Green Federal Building 1600 Arch Street Boston, MA 02109	Connecticut, Delaware, Maine, Massachusetts, New Hampshire, New Jersey, New York, Pennsylvania, Puerto Rico, Maryland, Rhode Island, and Vermont.
30th Floor 230 S. Dearborn St. Chicago, IL 60604	Illinois, Indiana, Iowa, Kansas, Kentucky, Michigan, Minnesota, Missouri, Nebraska, North Dakota, Ohio, South Dakota, West Virginia, and Wisconsin.

1100 Commerce Street Dallas, TX 75242	Arizona, Arkansas, Colorado, Louisiana, Montana, New Mexico, Oklahoma, Texas, Utah, and Wyoming.
7th Floor 211 Main Street San Francisco, CA 94105	Alaska, California, Hawaii, Idaho, Nevada, Oregon, Washington, and the Pacific Ocean Area.

APPENDIX C

STATE AND TERRITORIAL APPRENTICESHIP AGENCIES

ARIZONA
Apprenticeship Services
Department of Economic Security
438 W. Adams
Phoenix, AZ 85003

CALIFORNIA
Div. of Apprenticeship Standards
525 Golden Gate Avenue
P.O. Box 603
San Francisco, CA 94101

CONNECTICUT
Office of Job Training and Skill Dev., Labor Department
200 Folly Brook Boulevard
Wethersfield, CT 06109

DELAWARE
Delaware State Dept. of Labor
Appren. & Trng. Section
State Office Bldg., 6th Fl.
820 N. French Street
Wilmington, DE 19801

DISTRICT OF COLUMBIA
DC Apprenticeship Council
500 C Street, NW - Suite 241
Washington, DC 20001

FLORIDA
Division of Labor, Employment and Trng.
Bureau of Appren.
1320 Executive Center Drive
Atkins Bldg., Room 211
Tallahassee, FL 32399-0669

HAWAII
Apprenticeship Division
Dept. of Labor & Ind.
Relations - Room 202
830 Punch Bowl Street
Honolulu, HI 96813

KANSAS
Apprentice Training Section
Department of Human
Resources
512 W. 6th Street
Topeka, KS 66603

KENTUCKY
Apprenticeship and Training
Department of Labor
620 South Third Street
Louisville, KY 40202

LOUISIANA
Louisiana Dept. of Labor
Office of Labor
5360 Florida Boulevard
Baton Rouge, LA 70806

MAINE
Bureau of Apprenticeship
State ofc. Bldg.-7th Fl.
Station # 45
Augusta, ME 04333

MARYLAND
Apprenticeship & Trng.
Council
Division of Labor & Industry
5200 Westland Boulevard
Baltimore, MD 21227

MASSACHUSETTS
Dept. of Labor & Industries
Div. of Apprentice Training
Leverett Saltonstall Bldg.
100 Cambridge Street
Boston, MA 02202

MINNESOTA
Division of Apprenticeship
Dept. of Labor & Industry
Space Center Bldg., 4th Fl.
444 Lafayette Road
St. Paul, MN 55101

MONTANA
Apprenticeship Bureau
Labor Standards Division
Department of Labor &
Industry
Capitol Station-P.O. Box
1728
Helena, MT 59624

NEVADA
State Apprenticeship Council
505 East King Street, Room 602
Carson City, NE 98710

NEW HAMPSHIRE
Apprenticeship Council
19 Pillsbury Street
Concord, NH 03301

NEW MEXICO
State Apprenticeship Council
2340 Menaul Blvd., NE - Suite 212
Albuquerque, NM 87107

NEW YORK
Bureau of Employability Dev.
NY State Department of Labor
State Campus Bldg. 12, Room 111
Albany, NY 12240

NORTH CAROLINA
Apprenticeship Division
NC Department of Labor
Shore Memorial Bldg.
214 W. Jones Street
Raleigh, NC 27603

OHIO
Ohio State Apprenticeship Council
2323 W. Fifth Avenue, Room 2140
P.O. Box 825
Columbus, OH 43266-0567

OREGON
Apprenticeship & Trng. Div.
State Bureau of Labor & Industry
State Office Bldg.-Room 405
1400 S. West Fifth Avenue
Portland, OR 97201

PENNSYLVANIA
Apprenticeship and Training
Labor & Industry Bldg.
7th & Forster Street, Room 1303
Harrisburg, PA 17120

PUERTO RICO
Apprenticeship Division
Right to Employment Admin.
G.P.O. Box 4452
San Juan, PR 00936

RHODE ISLAND
RI State Apprenticeship Council
220 Elwood Avenue
Providence, RI 02907

VERMONT
Apprenticeship and Training
Department of Labor & Industry
120 State Street
Montpelier, VT 05602

VIRGIN ISLANDS
Division of Apprenticeship & Trng.
Department of Labor
P.O. Box 890
Christiansted, St. Croix, VI 00820

VIRGINIA
Division of Apprentice Training
Dept. of Labor & Industry
205 No. 4th Street-Room M-3
P.O. Box 12064
Richmond, VA 23241

WASHINGTON
Apprenticeship & Training Section
Department of Labor & Industries
925 Plum Street, Stop HC-710
Olympia, WA 98504

WISCONSIN
Bureau of Apprenticeship Standards
Dept. of Industry, Labor and Human Relations
201 E. Washington Ave., Room 211-X
P.O. Box 7972
Madison, WI 53707

APPENDIX D

FIELD AND REGIONAL OFFICES OF THE BUREAU OF APPRENTICESHIP TRAINING U.S. DEPARTMENT OF LABOR

REGION I

CONNECTICUT

Hartford	Room 367 Federal Building 135 High Street, 06103

MAINE

Augusta	Room 408-D Federal Building 68 Sewall Street, 04330

MASSACHUSETTS

Boston	11th Floor One Congress Street, 02114
Springfield	Room 211 Springfield Federal Building 1550 Main Street, 01103
Worcester	Room 316, Federal Building 595 Main Street, 01601

NEW HAMPSHIRE

Concord	143 North Main Street, 03301

RHODE ISLAND

Providence	Providence Federal Building 100 Hartford Avenue, 02909

VERMONT

Burlington	Suite 103 Burlington Square 96 College Street, 05401

REGION II

NEW JERSEY

Iselin	Parkway Towers Building E, Third Floor, 08830

NEW YORK

Albany	Leo O'Brien Federal Building North Pearl and Clinton Avenue, 12202
New York	Room 602, Federal Building 201 Varick Street, 10014
Buffalo	Room 209, Federal Building 111 West Huron Street, 14202
Rochester	100 State Street Federal Building, 14614
Syracuse	Federal Building 100 South Clinton Street, 13260

REGION III

DELAWARE

Wilmington	Federal Building 844 King Street, 19801

DISTRICT OF COLUMBIA

D.C.I.	Room 517 1111-20th Street, NW, 20036

MARYLAND

Baltimore	Room 1028 31 Hopkins Plaza Charles Center, 21201

PENNSYLVANIA

Erie	Room 106, Federal Building 6th and State Streets, 16507
Harrisburg	Federal Building 228 Walnut Street, 17108
Philadelphia	Room 13240 3535 Market Street, 19104
Pittsburg	Room 1436 Federal Building 1000 Liberty Avenue, 15222
Reading	Room 2115, East Shore Office Building 45 South Front Street, 19603
Wilkes-Barre	Room 2028, Penn Place 20 North Pennsylvania Avenue, 18701

VIRGINIA

Richmond	Room 10-020 400 N. 8th Street, 23240

WEST VIRGINIA

Clarksburg	Room 701 Palace Furniture Building 168 West Main Street, 26301

REGION IV

ALABAMA

Birmingham	Suite 102 Berry Building 2017 2nd Avenue, North, 35203

FLORIDA

Miami	995 N.W. 119th Street, 33168
Tallahassee	Suite 264 2574 Seagate Drive, 32301

GEORGIA

Atlanta	Suite 200 1371 Peachtree Street, NE, 30367
Columbus	Suite 24 3604 Macon Road, 31907
Savannah	Suite 303 120 Barnard Street, 31401

KENTUCKY

Louisville	Room 187-J Federal Building 600 Federal Place, 40202
Lexington	2033 Regency Road, 40503

MISSISSIPPI

Jackson	Suite 1010 Federal Building 101 West Capitol Street, 39269

NORTH CAROLINA

Raleigh	Suite 375 Somerset Center 4505 Falls of the Neuse Road, 27609

SOUTH CAROLINA

Columbia	Suite 838 Strom Thurmond Federal Bldg. 1835 Assembly Street, 29201
Charleston	Suite 313, Federal Building 334 Meeting Street, 29403

TENNESSEE

Nashville	Suite 101-A 460 Metroplex Drive, 37211

REGION V

ILLINOIS

Chicago	Room 758 230 S. Dearborn Street, 60604
Alton	501 Belle Street, 62002
Des Plaines	1420 Miner Street, 60016
Peoria	100 N.E. Monroe Street, 61602
Rockford	211 South Court Street, 61101
Springfield	U.S. Post Office, Room 14 600 East Monroe Street, 62701

INDIANA

Indianapolis	Room 414 Federal Building 46 East Ohio Street, 46204

MICHIGAN

Detroit	Room 657 231 W. Lafayette Avenue, 48226

MINNESOTA

St. Paul	Room 134 Federal Building 316 Robert Street, 55101
Duluth	Room 234 515 West First Street, 55802

OHIO

Cleveland	Suite 602 1375 Euclid Avenue, 44115

WISCONSIN

Madison	Room 303, Federal Center 212 East Washington Avenue, 53703

REGION VI

ARKANSAS

Little Rock	Room 3507 Federal Building 700 West Capitol Street, 72201

LOUISIANA

New Orleans	Room 1323, U.S. Postal Building 701 Loyola Street, 70115

NEW MEXICO

Albuquerque	Room 830 505 Marquette, 87102

OKLAHOMA

Tulsa	Room 305, 51 Yale Building 5110 South Yale, 74135

TEXAS

Dallas	Room 502, Federal Building 525 Griffin Street, 75202

REGION VII

IOWA

Des Moines	Federal Office Building 210 Walnut Street, 50309

KANSAS

Kansas City	Room 1100, Federal Office Building 911 Walnut Street, 64016

Wichita
Room B-41
U.S. Courthouse Building
401 North Market, 67202

MISSOURI

St. Louis
Robert A. Young Federal Building
1222 Spruce Street, 63103

NEBRASKA

Omaha
Room 801
106 South 15th Street, 68102

REGION VIII

COLORADO

Denver
Room 476
U.S. Custom House
721 19th Street, 80202

MONTANA

Helena
Room 394
Federal Office Building
301 South Park Avenue, 59626

NORTH DAKOTA

Fargo
Room 428
New Federal Building
653 2nd Avenue N., 58102

SOUTH DAKOTA

Sioux Falls	Room 107 Courthouse Plaza 300 North Dakota Avenue, 57102

UTAH

Salt Lake City	Room 1051 Administration Building 1745 West 1700 South, 84101

WYOMING

Cheyenne	Room 5013 Joseph C. Mahoney Federal Center 2120 Capitol Avenue, 82201

REGION IX

ARIZONA

Phoenix	3221 North 16th Street, 85016

CALIFORNIA

San Francisco	Suite 715 71 Stevenson Street, 94105

HAWAII

Honolulu	Room 5113 300 Ala Moana Boulevard, 96580

NEVADA

Las Vegas	Room 311, U.S. Court House 301 Stewart Avenue, 89101

REGION X

ALASKA

Anchorage	Room 554, Federal Building 222 W. 7th Street, 99513

IDAHO

Boise	Suite 128 3050 North Lakeharbor Lane, 83703

OREGON

Portland	526 Federal Building 1220 S.W. 3rd Avenue, 97204

WASHINGTON

Seattle	Room 925 1111 Third Avenue, 98101

APPENDIX E

SELECTED SCHOOLS OFFERING COURSES IN DRAFTING

Listed are schools throughout the United States where the drafter can find courses to further her or his education and technical competence. Space is too limited for a complete listing of all such sources. Represented therefore, are selected two-year community, technical, and junior colleges and post-high school technical institutes, all of which offer courses in some area useful to a drafter. Some four-year institutions are also listed.

In addition, many of these schools provide basic training in drafting skills for the individual who may wish to enter the field.

If the listing does not provide the leads you need, get in touch with two-year colleges or four-year colleges within reach, and find out their offerings. Also get in touch with your local board of education and your state's department of education, which usually is located in the capital city.

Alabama

Jefferson State Community College, Birmingham 35215

John C. Calhoun State Community College, Decatur 35609

Northwest Alabama Community College, Phil Campbell 35581

Shelton State Community College, Tuscaloosa 35404

George C. Wallace State Community College, Dothan 36303

Alaska

Matanuska-Susitna College, Palmer 99645

University of Alaska Southeast, Juneau 99801

Arizona

Arizona Western College, Yuma 85364

Central Arizona College, Coolidge 85228

Eastern Arizona College, Thatcher 85552

Mesa Community College, Mesa 85202

Phoenix College, Phoenix 85013

Pima Community College, Tucson 85702

Arkansas

East Arkansas Community College, Forrest City 72335

Garland County Community College, Hot Springs 71913

Westark Community College, Fort Smith 72913

California

American River College, Sacramento 95841
Antelope Valley College, East Lancaster 93534
Cerritos College, Norwalk 90650
Citrus College, Azusa 91702
City College of San Francisco, San Francisco 94112
College of Marin, Kentfield 94904
College of the Sequoias, Visalia 93277
Compton Community College, Compton 90221
Contra Costa College, San Pablo 94806
East Los Angeles College, Monterey Park, 91754
El Camino College, Via Torrance 90506
Foothill College, Los Altos Hills 94022
Fresno City College, Fresno 93741
Fullerton College, Fullerton 92634
Glendale Community College, Glendale 91208
Golden West College, Huntington Beach 92647
Hartnell College, Salinas 93901
Kings River Community College, Reedley 93654
Las Positas College, Livermore 94550
Long Beach City College, Long Beach 90808
Los Angeles City College, Los Angeles 90029
Los Angeles Harbor College, Wilmington 90744
Los Angeles Pierce College, Woodland Hills 91371
Los Angeles Trade-Technical College, Los Angeles 90015
Modesto Junior College, Modesto 95350
Monterey Peninsula College, Monterey 93940
Mount San Antonio College, Walnut 91789
Napa Valley College, Napa 94558
Orange Coast College, Costa Mesa 92628

Palomar College, San Marcos 92069
Pasadena City College, Pasadena 91106
Riverside Community College, Riverside 92506
Sacramento City College, Sacramento 95822
Saddleback College, Mission Viejo 92692
San Bernardino Valley College, San Bernardino 92410
San Diego City College, San Diego 92101
San Jose City College, San Jose 95128
Santa Barbara Community College, Santa Barbara 93109
Santa Monica College, Santa Monica 90405
Santa Rosa Junior College, Santa Rosa 95401
Shasta College, Redding 96099
Sierra College, Rocklin 95677
Solano Community College, Suisun City 94585
Taft College, Taft 93268
Yuba College, Maryville 95901

Colorado

Denver Institute of Technology, Denver 80221
Fort Lewis Agricultural & Mechanical College, Durango 81301
Mesa College, Grand Junction 81501
Northeastern Junior College, Sterling 80751
Otero Junior College, La Junta 81050
Trinidad State Junior College, Trinidad 81082

Connecticut

Hartford State Technical College, Hartford 06101
University of New Haven, West Haven 06516
Thames Valley State Technical College, Norwich 06360

District of Columbia

American University, Washington 20016
University of the District of Columbia, Washington 20008

Florida

Central Florida Community College, Ocala 32678
Daytona Beach Community College, Daytona Beach 32015
Edison Community College, Fort Myers 33906
Edward Waters College, Jacksonville 32209
Embry Riddle Aeronautical University, Daytona Beach 32114
Gulf Coast Community College, Panama City 32401
Miami-Dade Community College, Miami 33176
North Florida Junior College, Madison 32340
Palm Beach Community College, Lake Worth 33461
Pensacola Junior College, Pensacola 32504
Santa Fe Community College, Gainesville 32601
Seminole Community College, Sanford 32773

Georgia

Brunswick Junior College, Brunswick 31523
Dalton College, Dalton 30720
Darton College, Albany 31707
Middle Georgia College, Cochran 31014
Truett-McConnell College, Cleveland 30528

Hawaii

Honolulu Community College, Honolulu 96813
Leeward Community College, Pearl City 96782
Maui Community College, Kahului 96732

Idaho

College of Southern Idaho, Twin Falls 83301
North Idaho College, Coeur D'Alene 83814
Ricks College, Rexburg 83440

Illinois

Belleville Area College, Belleville 62221
Chicago City Colleges, Chicago 60601
Danville Area Community College, Danville 61832
Elgin Community College, Elgin 60120
Joliet Junior College, Joliet 60436
Kaskaskia College, Centralia 62801
Lake Land College, Mattoon 61938
Rend Lake College, Ina 62846
Triton College, River Grove 60171

Indiana

Indiana Vocational-Technical College, Kokomo 46901
Indiana Vocational-Technical College, South Bend 46619
Indiana Vocational-Technical College Northeast, Fort Wayne 46805
Vincennes University, Vincennes 47591

Iowa

Des Moines Area Community College, Ankeny 50021
Indian Hills Community College, Ottumwa 52501
Kirkwood Community College, Cedar Rapids 52406
Southwestern Community College, Creston 50801

Kansas

Coffeyville Community College, Coffeyville 67377
Cowley County Community College, Arkansas City 67005
Dodge City Community College, Dodge City 67801
Garden City Community Junior College, Garden City 67846
Highland Community College, Highland 66035
Hutchinson Community College, Hutchinson 67501
Independence Community College, Independence 67301
Johnson County Community College, Overland Park 66210
Kansas City Kansas Community College, Kansas City 66112
Pratt Community College, Pratt 67124

Kentucky

Lees College, Jackson 41339
University of Kentucky Community College System, Lexington 40506
Louisville Technical Institute, Louisville 40218

Louisiana

Bossier Parish Community College, Bossier City 71111
Delgado Community College, New Orleans 70119
Louisiana State University, Eunice 70535

Maine

Northern Maine Technical College, Presque Isle 04769
University of Maine, Augusta 04330

Maryland

Catonsville Community College, Catonsville 21228
Essex Community College, Baltimore 21237
Garrett Community College, McHenry 21541
Hagerstown Junior College, Hagerstown 21740
Harford Community College, Bel Air 21014
Montgomery College, Rockville 20850
Prince George's Community College, Largo 20772

Massachusetts

Massasoit Community College, Brockton 02402
Middlesex Community College, Bedford 01730
Springfield Technical Community College, Springfield 01105

Michigan

Charles Stewart Mott Community College, Flint 48502
Gogebic Community College, Ironwood 49938
Grand Rapids Junior College, Grand Rapids 49503
Henry Ford Community College, Dearborn 48128
Macomb County Community College, Warren 48093
Muskegon Community College, Muskegon 49442
Northwestern Michigan College, Traverse City 49684
Oakland Community College, Bloomfield Hills 48013
St. Clair County Community College, Port Huron 48060

Minnesota

Mesabi Community College, Virginia 55792
Rochester Technical College, Rochester 55904

Vermillion Community College, Ely 55731
St. Cloud Technical College, St. Cloud 56303

Mississippi

Copiah-Lincoln Community College, Wesson 39191
Hinds Community College, Raymond 39154
Itawamba Community College, Fulton 38843
Jones County Junior College, Ellisville 39437
Meridian Community College, Meridian 39305
Mississippi Gulf Coast Community College, Perkinston 39573
Northeast Mississippi Community College, Booneville 38829
Pearl River Community College, Poplarville 39470
Southwest Mississippi Community College, Summit 39666

Missouri

Crowder College, Neosho 64850
East Central College, Union 63084
Jefferson College, Hillsboro 63050
Longview Community College, Lee's Summit 64063
Missouri Western State College, St. Joseph 64507
Penn Valley Community College, Kansas City 64111
St. Louis Community College, St. Louis 63102

Montana

Dull Knife Memorial College, Lame Deer 59043
Billings Vocational-Technical Center, Billings 59102
Butte Vocational Technical Center, Butte 59701

Nebraska

Mid-Plains Community College, North Platte 69101
Northeast Community College, 68701

Nevada

Northern Nevada Community College, Elko 89801
Western Nevada Community College, Carson City 89701

New Hampshire

New Hampshire Technical College, Manchester 03104
New Hampshire Technical College, Stratham 03885

New Jersey

Bergen Community College, Peramus 07652
Gloucester County College, Sewell 08080
Salem Community College, Carneys Point 08069
Raritan Valley Community College, Somerville 08876

New Mexico

Eastern New Mexico University, Portales 88130
New Mexico Junior College, Hobbs 88240
Northern New Mexico Community College, Espanola 87532

New York

Adirondack Community College, Queensbury 12804
Broome Community College, Binghamton 13902
Columbia-Greene Community College, Hudson 12534
Corning Community College, Corning 14830
Dutchess Community College, Poughkeepsie 12601

Genesee Community College, Batavia 14020
Herkimer County Community College, Herkimer 13350
Erie Community College, Buffalo 14203
Hudson Valley Community College, Troy 12180
Mohawk Valley Community College, Utica 13501
New York Institute of Technology, Old Westbury 11568
Orange County Community College, Middletown 10940
Queensborough Community College, Bayside, New York 11364
Rockland County Community College, Suffren 10901
Westchester Community College, Valhalla 10595
Ulster County Community College, Stone Ridge 12484

North Carolina

Asheville-Buncombe Technical College, Asheville 28801
Beaufort County Community College, Washington 27889
Catawba Valley Community College, Hickory 28601
College of the Albermarle, Elizabeth City 27909
Western Piedmont Community College, Morganton 28655
Wingate College, Wingate 28174
Wilkes Community College, Wilkesboro 28697

North Dakota

North Dakota State School of Science, Wahpeton 58075
University of North Dakota, Williston 58801

Ohio

Belmont Technical College, St. Clairsville 43950
Clark State Community College, Springfield 45505
Columbus State Community College, Columbus 43216
Sinclair Community College, Dayton 45402
University of Toledo, Toledo 43606

Oklahoma

Cameron University, Lawton 73501
Eastern Oklahoma State College, Willburton 74578
Murray State College, Tishomingo 73460
Northeastern Oklahoma Agricultural & Mechanical College, Miami 74354
Rogers State College, Claremore 74017
Western Oklahoma State College, Altus 73521

Oregon

Central Oregon Community College, Bend 97701
Linn-Benton Community College, Albany 97321
Oregon Institute of Technology, Klamath Falls 97601
Umpqua Community College, Roseburg 97470
Treasure Valley Community College, Ontario 97914

Pennsylvania

Bucks County Community College, Newtown 18940
Butler County Community College, Butler 16001

Community College of Beaver County, Monaca 15061
Harrisburg Area Community College, Harrisburg 17110
Lehigh County Community College, Schnecksville 18078
Luzerne County Community College, Nanticoke 18634
Pennsylvania State University, Commonwealth Campuses, University Park 16802

South Carolina

Greenville Technical College, Greenville 29606
Midlands Technical College, Columbia 29250
Tri-County Technical College, Pendleton 29670
Trident Technical College, Charleston 29406

South Dakota

Northern State University, Aberdeen 57401

Tennessee

Cleveland State Community College, Cleveland 37320
Jackson State Community College, Jackson 38301
Nashville State Technical Institute, Nashville 37209
Tennessee Technological University, Cookeville 38501

Texas

Amarillo College, Amarillo 79178
Central Texas College, Killeen 76541
Del Mar College, Corpus Christi 78404
Frank Phillips College, Borger 79008

Henderson County Junior College, Athens 75751
Howard College, Big Spring 79720
Kilgore College, Kilgore 75662
Laredo Junior College, Laredo 78040
Lee College, Baytown 77520
Navarro College, Corsicana 75110
Odessa College, Odessa 79762
Richland College, Dallas 75243
St. Phillips College, San Antonio 78203
South Plains College, Levelland 79336
Southwest Texas Junior College, Uvalde 78801
Temple Junior College, Temple 76501
Texarkana College, Texarkana 75501
Victoria College, Victoria 77901
Weatherford College, Weatherford 76086
Wharton County Junior College, Wharton 77488

Utah

Salt Lake Community College, Salt Lake City 84130
Utah Valley Community College, Orem 84058
Weber State College, Ogden 84408

Vermont

Community College of Vermont, Montpelier 05602
Vermont Technical College, Randolph Center 05061

Virginia

Blue Ridge Community College, Weyers Cave 24486
Dabney S. Lancaster Community College, Clifton Forge 24422
Germanna Community College, Locust Grove 22508

New River Community College, Dublin 24084
Piedmont Virginia Community College, Charlottesville 22901
Tidewater Community College, Portsmouth 23703

Washington

Centralia College, Centralia 98531
Clark Community College, Vancouver 98663
Columbia Basin College, Pasco 99301
Everett Community College, Everett 98201
Lower Columbia College, Longview 98632
Olympic College, Bremerton 98310
Skagit Valley College, Mount Vernon 98273
Yakima Valley Community College, Yakima 98907

West Virginia

Southern West Virginia Community College, Logan 25601
West Virginia Institute of Technology, Montgomery 25136
West Virginia Northern Community College, Wheeling 26003
West Virginia University of Parkersburg, Parkersburg 26101

Wisconsin

Black Hawk Technical College, Janesville 53547
Gateway Technical College, Kenosha 53141
Madison Area Technical College, Madison 53703
Milwaukee Area Technical College, Milwaukee 53203

Wyoming

Casper College, Casper 82601

Central Wyoming College, Riverton 82501

Sheridan College, Sheridan 82801

Western Wyoming Community College, Rock Springs 82902

VGM CAREER BOOKS/CAREERS FOR YOU

OPPORTUNITIES IN
Accounting
Acting
Advertising
Aerospace
Agriculture
Airline
Animal and Pet Care
Architecture
Automotive Service
Banking
Beauty Culture
Biological Sciences
Biotechnology
Book Publishing
Broadcasting
Building Construction Trades
Business Communication
Business Management
Cable Television
CAD/CAM
Carpentry
Chemistry
Child Care
Chiropractic
Civil Engineering
Cleaning Service
Commercial Art and Graphic Design
Computer Maintenance
Computer Science
Counseling & Development
Crafts
Culinary
Customer Service
Data Processing
Dental Care
Desktop Publishing
Direct Marketing
Drafting
Electrical Trades
Electronic and Electrical Engineering
Electronics
Energy
Engineering
Engineering Technology
Environmental
Eye Care
Fashion
Fast Food
Federal Government
Film
Financial
Fire Protection Services
Fitness
Food Services
Foreign Language
Forestry
Government Service
Health and Medical
High Tech
Home Economics
Homecare Services
Hospital Administration
Hotel & Motel Management
Human Resources Management Careers
Information Systems
Insurance
Interior Design
International Business
Journalism
Laser Technology
Law
Law Enforcement and Criminal Justice
Library and Information Science
Machine Trades
Magazine Publishing
Marine & Maritime
Masonry
Marketing
Materials Science
Mechanical Engineering
Medical Imaging
Medical Technology
Metalworking
Microelectronics
Military
Modeling
Music
Newspaper Publishing
Nonprofit Organizations
Nursing
Nutrition
Occupational Therapy
Office Occupations
Packaging Science
Paralegal Careers
Paramedical Careers
Part-time & Summer Jobs
Performing Arts
Petroleum
Pharmacy
Photography
Physical Therapy
Physician
Plastics
Plumbing & Pipe Fitting
Postal Service
Printing
Property Management
Psychology
Public Health
Public Relations
Purchasing
Real Estate
Recreation and Leisure
Refrigeration and Air Conditioning
Religious Service
Restaurant
Retailing
Robotics
Sales
Secretarial
Securities
Social Science
Social Work
Speech-Language Pathology
Sports & Athletics
Sports Medicine
State and Local Government
Teaching
Technical Communications
Telecommunications
Television and Video
Theatrical Design & Production
Tool and Die
Transportation
Travel
Trucking
Veterinary Medicine
Visual Arts
Vocational and Technical
Warehousing
Waste Management
Welding
Word Processing
Writing
Your Own Service Business

CAREERS IN Accounting; Advertising; Business; Communications; Computers; Education; Engineering; Finance; Health Care; High Tech; Law; Marketing; Medicine; Science; Social and Rehabilitation Services

CAREER DIRECTORIES
Careers Encyclopedia
Dictionary of Occupational Titles
Occupational Outlook Handbook

CAREER PLANNING
Admissions Guide to Selective Business Schools
Beginning Entrepreneur
Career Planning and Development for College Students and Recent Graduates
Careers Checklists
Careers for Animal Lovers
Careers for Bookworms
Careers for Computer Buffs
Careers for Crafty People
Careers for Culture Lovers
Careers for Environmental Types
Careers for Film Buffs
Careers for Foreign Language Aficionados
Careers for Good Samaritans
Careers for Gourmets
Careers for Nature Lovers
Careers for Numbers Crunchers
Careers for Sport Nuts
Careers for Travel Buffs
Cover Letters They Don't Forget
Guide to Basic Resume Writing
How to Approach an Advertising Agency and Walk Away with the Job You Want
How to Bounce Back Quickly After Losing Your Job
How to Change Your Career
How to Choose the Right Career
How to Get and Keep Your First Job
How to Get into the Right Law School
How to Get People to Do Things Your Way
How to Have a Winning Job Interview
How to Jump Start a Stalled Career
How to Land a Better Job
How to Launch Your Career in TV News
How to Make the Right Career Moves
How to Market Your College Degree
How to Move from College into a Secure Job
How to Negotiate the Raise You Deserve
How to Prepare a *Curriculum Vitae*
How to Prepare for College
How to Run Your Own Home Business
How to Succeed in College
How to Succeed in High School
How to Write a Winning Resume
How to Write Your College Application Essay
Joyce Lain Kennedy's Career Book
Resumes for Advertising Careers
Resumes for Banking and Financial Careers
Resumes for College Students & Recent Graduates
Resumes for Communications Careers
Resumes for Education Careers
Resumes for Health and Medical Careers
Resumes for High School Graduates
Resumes for High Tech Careers
Resumes for Midcareer Job Changes
Resumes for Sales and Marketing Careers
Resumes for Scientific and Technical Careers
Successful Interviewing for College Seniors

VGM Career Horizons
a division of *NTC Publishing Group*
4255 West Touhy Avenue
Lincolnwood, Illinois 60646-1975